MANCHESTER SHIP CANAL

EDWARD GRAY

SUTTON PUBLISHING LIMITED

Sutton Publishing Limited
Phoenix Mill · Thrupp · Stroud
Gloucestershire · GL5 2BU

First published 1997

British Library Cataloguing in Publication Data
A catalogue record for this book is available from the British Library.

ISBN 0-7509-1459-9

Typeset in 10/12 Perpetua.
Typesetting and origination by
Sutton Publishing Limited.
Printed in Great Britain by
Ebenezer Baylis, Worcester.

The Manchester Ship Canal's main terminal docks at Salford in 1966. The expanse of water in the foreground was the turning basin, which allowed large ships to be swung round for their return journey along the canal. (Airviews)

CONTENTS

Manchester Liners' container ship *Manchester Challenge* passes along the Ship Canal near Warburton Bridge in 1975.

INTRODUCTION

The construction of the Manchester Ship Canal was hailed as the greatest engineering achievement of Victorian times. It transformed an inland city into a major seaport. In 1882 dissatisfaction with high railway rates and oppressive dock dues charged on goods passing through the Port of Liverpool inspired a group of businessmen to attend a meeting at the Didsbury home of Daniel Adamson, a Durham-born Manchester industrialist. As a result, a detailed survey was commissioned to examine the feasibility of a canal that would be wide and deep enough to enable ocean-going ships to load and discharge goods in the heart of the south-east Lancashire industrial area. The success of the 1869 Suez Canal project offered encouragement, and it was resolved that any new canal would be of similar depth and width, so that any ship which could sail through Suez would be able to reach Manchester.

The 1882 meeting was not the first to consider proposals to bring ships to Manchester. A scheme of 1660 to dredge and widen the Rivers Mersey and Irwell and so make them more easily navigable, was revived in 1712 when Thomas Steers published his recommendations for a waterway from the Mersey near Warrington as far as Hunt's Bank in Manchester. Steers proposed to incorporate a tow-path, so that men or horses could haul the boats, and to install sets of locks and weirs to ensure a good depth of water and lift the boats to Manchester's height above sea level. The Mersey & Irwell Navigation Company was formed to obtain the enabling Act in 1721, but the start was delayed and progress slow because of the shallow and winding nature of the rivers. By 1734 ships of moderate size were able to reach Manchester, and the importance of the cheap carriage of raw materials and manufactured goods became obvious as business prospered. The first section of the Duke of Bridgewater's canal, from Worsley to Manchester, followed in 1761 and greatly reduced the cost of coal in the city centre.

The river navigation was slow and tortuous, so when the Duke extended his canal to Runcorn in 1776 an alternative and more reliable route to the sea became available. The Mersey & Irwell Company countered by constructing an 8-mile canal from Latchford to Runcorn where a new lock allowed vessels to enter the Mersey at a deeper part of the estuary. Further inland, additional 'cuts' were made across loops of the river to shorten the sailing distance to Manchester. The river and canal companies then agreed to charge identical tolls to end ruinous competition, and both remained busy. Sailing barges or river 'flats' tended to tranship their cargoes at Runcorn or cross to Liverpool by sail, assisted by the tidal flow. But both waterways could accommodate vessels of only limited size, and in 1824 Matthew Hedley led an abortive scheme for a large canal from the Dee estuary.

The opening of the Liverpool & Manchester Railway in 1830 did not immediately affect traffic on the waterways. In 1838 canal engineer Sir John Rennie proposed a scheme to dredge a deep channel up the Mersey Estuary as far as Warrington, from where it could be extended to Manchester. In 1840 further schemes were advanced, one of which sought to enlarge and improve the existing Mersey & Irwell route. But this was the period of 'railway mania', when investors were more interested in extending the rail network. The river navigation began to suffer badly from railway competition, but the Bridgewater Canal had the advantage of connections to other canal systems, and remained more prosperous. In 1844 the Bridgewater Trustees bought out their old rival to become the owners of both waterways. From then on it was evident that maintenance was concentrated on the canal, for the river navigation, once able to accommodate boats with a draught of over 5 feet, by 1860 had difficulty in coping with vessels drawing only 3 feet. Problems on the river route were caused by flood damage, shifting mud-banks, the silting up of locks, and sometimes a shortage of water. By contrast, the Bridgewater Canal remained a keen competitor to the railways, carrying bulk cargoes at cheaper rates. In 1872 the Bridgewater Navigation Company, consisting mainly of railway shareholders, was formed to acquire and control both the canal and river navigation. Edward Leader Williams was appointed General Manager and Engineer.

The depression of the late 1870s found Manchester facing economic stagnation. Traders complained about the excessively high prices demanded by the railway monopoly, and the exorbitant rates charged on goods passing through Liverpool. Lower transport costs appeared to be necessary for business survival. It was cheaper to import goods via east-coast ports and pay railway charges to bring them across the Pennines, than to bring them in via Liverpool. Over half the cost of sending a ton of cotton goods to India was absorbed in railway and dock charges through Liverpool.

The advantages of a deep, wide waterway to link Manchester with the sea seemed very desirable. In 1877, and again in 1881, Hamilton Fulton's scheme for a tidal canal was considered, although it had the obvious disadvantage that the terminal docks would have had to have been at the base of a deep excavation. As a result of the 1882 meeting, the Ship Canal scheme was born. Fulton and Leader Williams (the latter by then an independent consultant) were each invited to report on the tidal navigation plan. Both favoured the idea of a dredged channel along the Mersey Estuary, but Williams' view, which found favour when he was appointed Engineer, was that the tidal flow would be feasible for only part of the course, beyond which locks would be needed.

Support for the Ship Canal grew, but opposition from railway companies and the vested interests at Liverpool dampened enthusiasm and made fund-raising difficult. Even so, a scheme based on the use of channels and training walls in the Mersey Estuary as far as Runcorn was prepared and submitted for Parliamentary approval in 1883. It was passed in the Commons, but rejected by the Lords. A second application in 1884 was passed by the Lords and rejected in the Commons. This time, some opponents of the bill revealed that they did not think Liverpool's trade would be harmed by a ship canal to Manchester, but they wished to be assured that any work carried out in the estuary would not damage it or cause the Liverpool bar to silt up. Consequently, the Bill was recast, and the idea of channels through the middle of the estuary was abandoned in favour of an extension of the canal along the southern shore.

The Manchester Ship Canal Act received Royal Assent in 1885. It incorporated the proviso that the Bridgewater Navigation, which included the river system, must be the subject of a compulsory purchase, and that the Bridgewater Canal should be maintained in good order and available for use. Over £500,000 had been spent in gaining Parliamentary approval. The next step for the promoters was to raise an estimated £6–7,000,000 before work could commence. Unhappily, differences arose between the promoters as to the best method of raising capital. Daniel Adamson resigned as chairman because of his firm belief that local enterprise should not be financed by London banks. Two years passed before sufficient funds were accumulated. In July 1887 the Bridgewater Navigation was purchased for £1,710,000, the largest amount written on a cheque to that date, and another £5,000,000 was made available to enable work to commence in November 1887. Over five years had passed between the 1882 meeting and the cutting of the first sod near Eastham. Sadly, Daniel Adamson, without whose determination the canal would probably have not been made, did not live to see its completion. He died in January 1890, aged 71. The full length of the canal was opened to traffic on New Year's Day 1894, the formal opening by Queen Victoria following on 21 May of that year.

The total expenditure on the Ship Canal to 1894 was £14,347,891, almost double the original estimate. But the canal cheapened the carriage of goods, attracted new industries, revitalized the economy of the region, and turned Manchester into a major port. In 1894 it handled 925,659 tons of cargo, and showed steady growth. The peak year for tonnage was 1959, with a grand total of 18,558,210.

SECTION ONE

CONSTRUCTION, 1887–1893

Britain's earliest canals had been constructed by the manual labour of large gangs of navvies, but by 1887, when work began on the Manchester Ship Canal, the old traditional methods had been largely superseded. The contractor, Thomas A.Walker, believed in mechanization, and consequently invested in a great many steam-powered shovels and land excavators. However, mechanical diggers were not always suitable for hard rock, or, as here at Acton Grange, where the cutting had to be made through an area of heavy clay. In such districts, the traditional pick-and-shovel methods survived. Spoil was loaded into railway wagons and hauled away to form embankments or to be dumped on low-lying grounds elsewhere. At the seaward end of the excavation, spoil dumped on the Pool Hall Rocks grew to such an extent that it was named 'Mount Manisty' after Mr E. Manisty, the contractor's agent for the Eastham section. (MSCCo)

The Mersey & Irwell Navigation Company had deepened the rivers and constructed locks in order to enable boats of a moderate size to reach Manchester as early as 1734. At Barton, near Eccles, boats using the river passed beneath this stone road bridge. Immediately beyond it was the 1761 aqueduct designed by James Brindley, which enabled barges on the rival Bridgewater Canal to cross the river navigation. A support of the aqueduct may be seen through the central arch of the bridge.

When the Manchester Ship Canal Bill finally received Parliamentary approval, a great demonstration was held in Eccles, where local residents believed it would bring prosperity to the district. A feature of the celebration was the roasting of an ox on 31 August 1885. Daniel Adamson himself attended and made the first ceremonial cut in the ox.

Mode Wheel weir, Mersey & Irwell Navigation. When the 1885 Ship Canal Act was approved, an estimated £6,000,000 had to be raised before work could begin. In 1887 £1,710,000 was expended in the compulsory purchase of the Bridgewater Navigation, which since 1872 had included the river system. A corn mill, powered by a wheel (Maud's Wheel) driven by the fall of water over the weir, had existed there since medieval times.

The weirs impeded river navigation, so the Mersey & Irwell Company constructed locks to raise and lower boats around the obstructions. Mode Wheel Lock, at the side of the mill, was a little over 15 feet wide. The arched bridge (right) led to a mooring basin. Mill and weir were destroyed when canal construction began. The size of boats using the river navigation was limited by the small 70-feet-long Calamanco Lock near Irlam.

Barton locks and weir on the old river navigation were located close to Brindley's Bridgewater Canal aqueduct. Here, too, was a corn mill, sited on the Trafford Park bank, on the right in this 1888 view looking upstream. By this date, very few craft used the neglected river navigation, which suffered from droughts, floods, shallows and accumulated debris.

A short distance upstream from the previous view, the planned course of the canal is marked out through the fields ahead. Bends in the river were straightened out during the construction, which necessitated some transfer of land from one bank to the other. Excavation of sections was carried out 'in the dry', before river water was turned in. The old river bed could then be used as a dump for spoil.

At the seaward end of the workings, different techniques were employed. Between Runcorn and Eastham over 13,000 piles were sunk into the river bed to support a 10-mile embankment which separated and sealed off the tidal estuary. A temporary dam (centre, left), built across the intended course of the canal near Runcorn Bridge, carried railway track to enable trains to dump material on the embankment. The small building (left) was formerly the public baths and starting point for the ferry to Widnes.

To transport men and equipment and to remove spoil from the cuttings, the contractor laid many miles of temporary railway lines alongside and in the base of the excavations. Some spoil was dumped on low-lying land, some was used to build up embankments where required. In several places the lines ran on temporary embankments across fields.

The use of so many short-lived and unstable earth embankments often led to minor mishaps, as seen here when the contractor's locomotive *Walmer* (a Manning Wardle product of 1878), became derailed on a collapsed portion of track. Much spoil was dumped on low-lying areas, and track was moved about as required from day to day.

Many locomotives brought to the canal construction had previously been employed on work elsewhere. The contractor also purchased over 100 new ones. *Raglan*, heading a spoil train at the base of a cutting where work is well advanced, was a Hunslet engine of 1887. It had been used first on Walker's Barry Docks contract, and was one of several retained afterwards for use on the Canal Company's system. (MSCCo)

The Mersey & Irwell Navigation system was irrevocably destroyed in the construction of the Ship Canal. A steam crane removes the lock gate at Barton old locks on 13 March 1891. Behind the lock-keeper's house (right) were stables for the horses employed in hauling the river barges. In the background looms the shape of Brindley's 1761 aqueduct, left intact for the time being to enable use of the Bridgewater Canal to continue uninterrupted.

The railway lines laid on the upper portions of the cuttings tended to be somewhat more permanent, and were used not only by contractor's trains, but also by the giant land-excavators as they moved their way along the workings. Traffic on the contractor's railway was so intense in certain busy sections that a signalling system was installed.

Wherever there were established crossing points along the projected course of the new canal, the contractor was obliged to institute alternative arrangements as the excavations grew deeper. Temporary bridges constructed of timber, such as this linking to Trafford Park, became common sights along the canal workings.

A large work-force, such as that employed on the canal construction, invariably attracted others offering services. Suppliers of food, drink, and clothing catered for the navvies. This 'coffee man' was photographed at Salford in July 1890. He carries a yoke bearing two stone jars, one of which is festooned with tin mugs. (MSCCo)

Little Bolton cutting, near Mode Wheel, May 1890, showing well-advanced excavation work. At this stage, the river flowed in its original course behind the embankment to the left. In some other sections, the river was diverted behind temporary embankments of earth known as 'coffer dams'. Water was turned into this section in 1891.

To maintain tools in good order and to provide servicing facilities, a number of temporary workshops were established along the projected course of the canal. This wooden hut was equipped with a workbench, dynamo, engine and lathe, plus cans of assorted lubricants. An enamel plate beneath the cans advertises Windermere steam launches.

Bridges, locks, and other major installations along the canal required substantial foundations at the base of the excavation. Here, stone blocks and brickwork are prepared prior to concreting work. The depth of the new cutting may be gauged from the height of the bank on the right.

The construction of the five sets of locks was a massive undertaking. The excavation at Irlam, looking eastwards to Manchester, shows the width of the canal at the entrance to the locks. The large lock chamber is to the left, the smaller in the centre, and the sluices are on the right. Railway tracks occupy the bed of the cutting. In the distance is a temporary bridge. Compare this scene with the bottom picture on page 74.

Barton, September 1891. The old stone road bridge was demolished and replaced with a temporary structure. Part of the steelwork for the new swing aqueduct is in place. The central pier, on which the two swing-bridges will turn, appears complete. The 'coffer dam' in the centre carries a railway line and separates work in progress from the river flowing on the right.

The stone aqueduct carrying the Bridgewater Canal over the Irwell remained in place until the last possible moment, so that barge traffic could continue unhindered. The diversion to the new swing aqueduct was prepared (left), though sealed off until work was complete. The first admission of water caused a portion of the approach to collapse, but a second attempt succeeded.

Until the new swing aqueduct came into use in August 1893, the demolition of Brindley's 1761 structure could not begin. The end of the new aqueduct may be noted (left), while in the foreground work is nearing completion on the destruction of the old stone aqueduct. Debris is being carried away in tubs on the river barge. (Trafford Libraries)

The new aqueduct was a huge steel tank, sealed at both ends, which could be swung while remaining full of water, so that no time was wasted in emptying or refilling. Barges cross on the Bridgewater Canal, the tow-path being the raised walkway on the left. The nearest barge is at the point where double gates seal both the aqueduct and the ends of the canal.

SECTION TWO

THE EARLY YEARS

The entrance to the Manchester Ship Canal at Eastham is approached via a dredged channel in the Mersey Estuary. The locks could be entered four hours before and after the time of peak tide, with priority being given to outgoing ships before high water. The sluice gates, which control the level of water in the section of the canal up to the next locks at Latchford, are to the right. The three lock chambers are of varying sizes in order to conserve water. The largest measures 600 by 80 feet, the medium 350 by 50 feet, and the smallest, intended for barges only and now disused, 150 by 30 feet. The railway-type signals (left) indicated to ships which lock should be used. The chosen depth for the canal was a minimum of 26 feet (as in the Suez Canal), but the lock sills were constructed 2 feet lower, enabling the depth to be increased to 28 feet in 1909. (F. Walker)

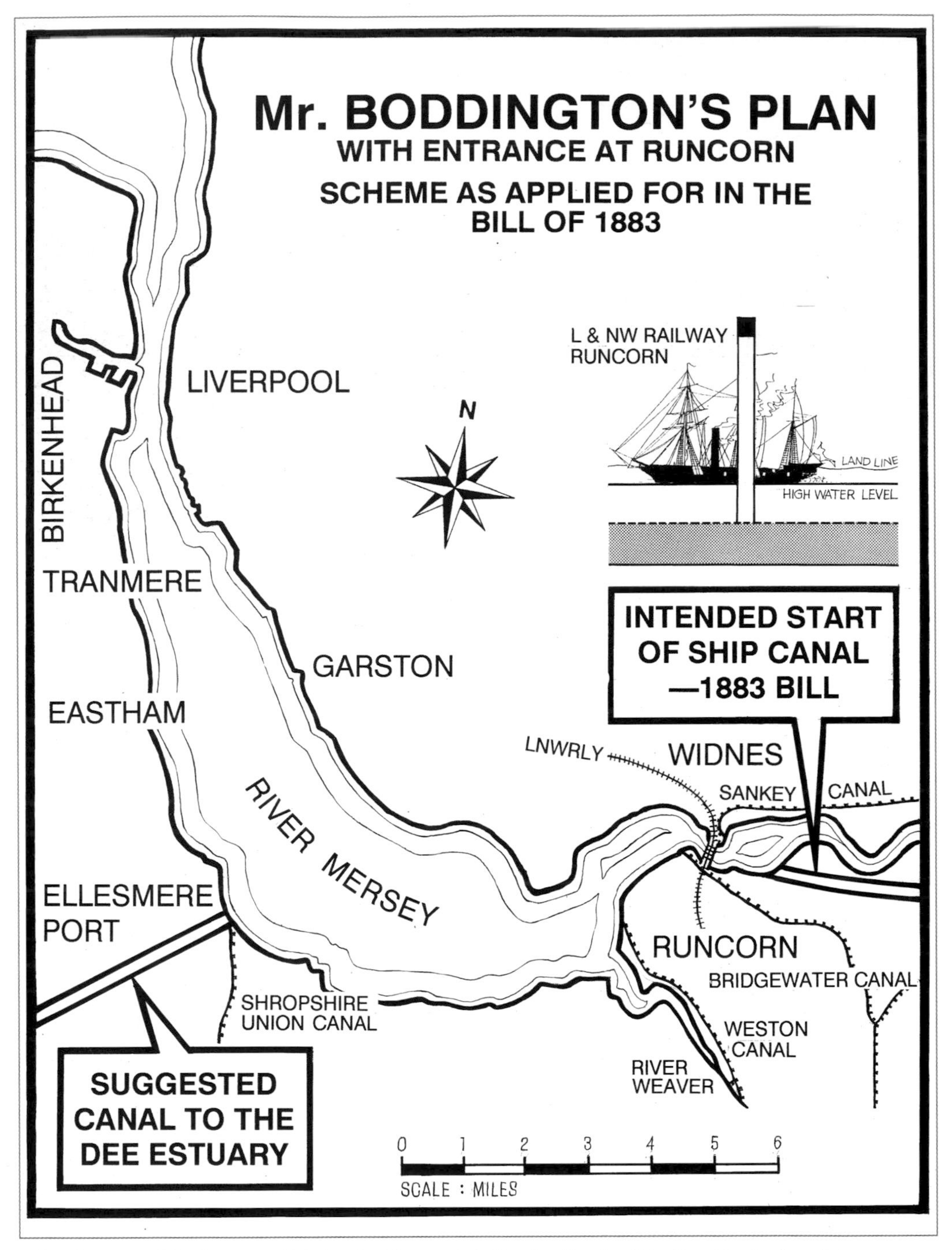

Fulton's 1882 tidal scheme and Henry Boddington's 1883 plan for the Ship Canal both placed the entrance at Runcorn, for it was not thought wise to obstruct existing commercial outlets to the Mersey Estuary from the Shropshire Union Canal, the Weaver Navigation, the Weston Canal, and the Bridgewater Canal. The proposed canal to the Dee Estuary was never constructed. (Alan Palmer)

The lighthouse, Ellesmere Port, built in about 1802, was formerly on the shoreline at the point where boats could enter the Shropshire Union Canal from the tidal waters of the Mersey. In 1892 a line of railway trucks (left) is dumping spoil on the newly built embankment separating the Ship Canal from the estuary. The lighthouse still exists, and is now part of the Boat Museum.

The western end of the canal was the first section to come into use. In order to minimize disruption and allow vessels continued access to the western ports, openings were left in the embankment which was to seal off the canal from the tidal waters of the Mersey. Efforts were concentrated on opening the section as far as the River Weaver, for until this was done, an Act of Parliament prevented work on the next section to Runcorn. The final gap was sealed at Ellesmere Port in July 1891, after which date ships entered via Eastham Locks. Vessels of up to 400 tons, small enough to have entered the port previously, were granted free use of the canal as far as Ellesmere Port, but ships of a greater size were charged a toll. The Pontoon Dock at Ellesmere Port, here occupied by the sailing ship *Stefano Razeto*, was provided with north-eastern finance in 1893. East-coast businessmen were more inclined to invest in the Ship Canal venture than west-coast shippers, who feared the canal might take away trade from their home ports.(W. Hall)

An 11-mile length of the canal was in use by September 1891, and temporary docks were constructed close to the mouth of the River Weaver for the import of grain and timber and the export of Cheshire salt. The chairman of the salt-exporters federation ceremonially named the wharf 'Saltport' in July 1892.

In June 1893 Runcorn Docks, which had been closed for twelve months, reopened. Here, the old docks are filled with sailing ships. Vessels which could not lower their masts to pass under the fixed bridges further along the Ship Canal could transfer their cargoes to lighters at Runcorn, which was also the transhipment point for goods arriving on the Bridgewater Canal. The first of the flight of locks leading to the Bridgewater Canal is in the foreground.

Edward Leader Williams, Engineer to the Ship Canal, had earlier been appointed to the Bridgewater Navigation in 1872, and in his new post had invested in a fleet of small steam-tugs to increase efficiency on the narrow-boat canal. Lacking a steady flow of water, the Bridgewater tended to freeze in severe weather conditions. Here, tugs attempt to break the ice, while a horse struggles on the tow-path.

The Bridgewater Company had purchased larger steam-tugs at an earlier date, but for use only on the Mersey Estuary, where they were used to tow barges across the river from Runcorn. Use of steam-powered tugs on the narrow canal had been avoided for fear that their wash might damage the banks. The 1863 tug *Dagmar* was one of nine larger tugs taken over on the purchase of the Bridgewater Navigation. (MSCCo)

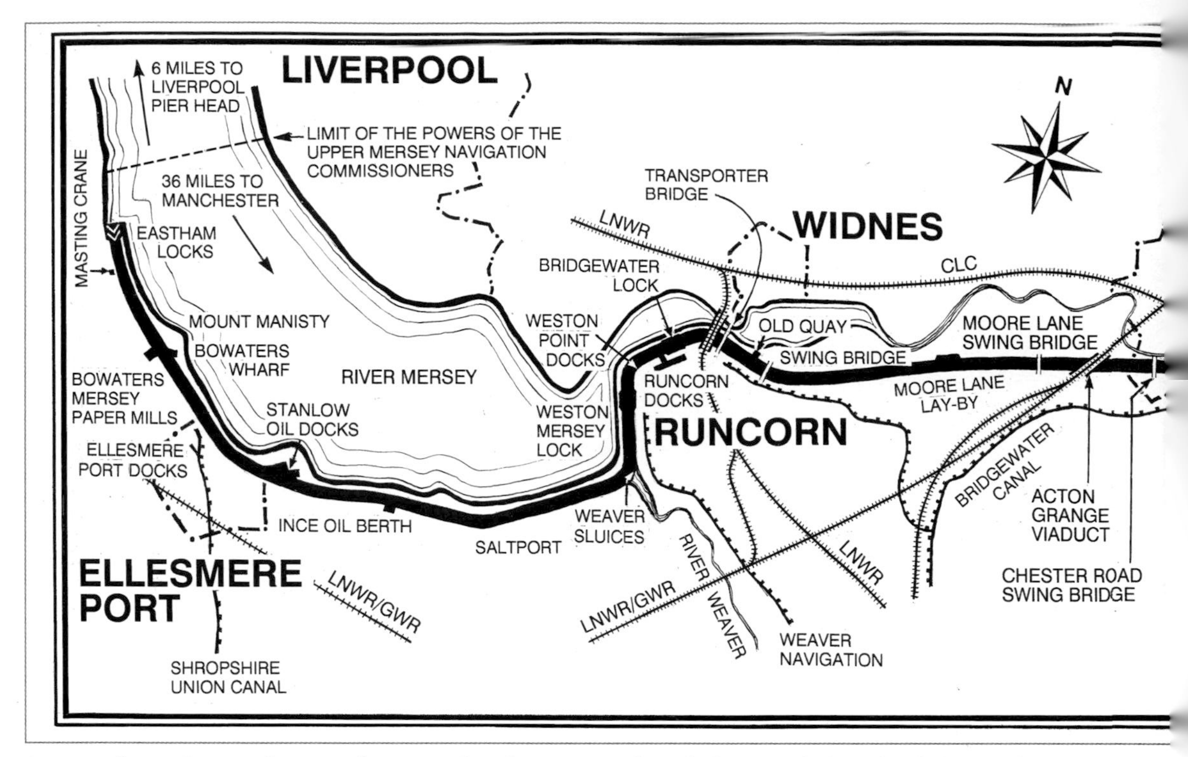

A map of Manchester Ship Canal as completed in 1894. The whole 36-mile length of the canal is designated the 'Port of Manchester'. (Alan Palmer)

New Year's Day 1894 saw the ceremonial opening of the full length of the canal. The directors had earlier embarked on a trial trip in December, and on 1 January a procession of seventy-one ships sailed from Latchford to the terminal docks. The Mersey ferryboat *Crocus*, carrying wives and friends of the directors, was the third in line as she passed through Barton. Note the sealed ends of the aqueduct and Bridgewater Canal. (R. Banks)

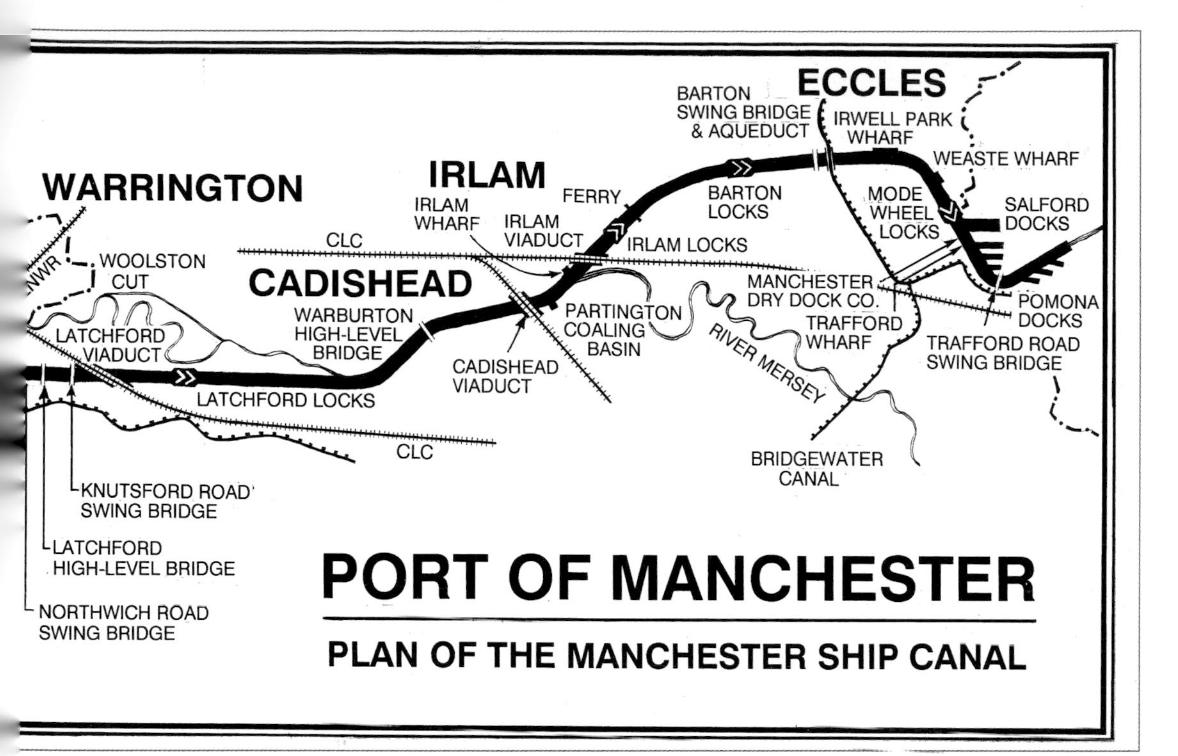

The first vessel to unload its cargo on the opening day was the Co-Operative Wholesale Society's ship *Pioneer,* which was also the first vessel registered in the Port of Manchester. The CWS instituted a regular weekly service to Rouen. The *Fraternity*, moored in Pomona Docks, displays unusual funnel markings. The CWS funnels carried the initials of the ship's name: 'F' for *Fraternity*, 'P' for *Pioneer* and 'NP' for *New Pioneer*. (A.Thirsk)

Timber and grain were quickly established as major imports. The steamship *Agenoria*, a 1883-built vessel owned by Rickenson Sons & Company of Hartlepool, discharges timber by the railway tracks on Trafford Wharf. Liverpool buyers attempted to impose a boycott on timber brought in via the Ship Canal. (Charles Downs)

The Norwegian steamship *Vafos* of Kristiania carries a deck cargo of timber as she leaves Mode Wheel locks heading for a berth in the terminal docks. It was common for timber ships to increase their carrying capacity by stacking timber between props on deck. The Harrison Line ship, *Musician* of Liverpool, sits high in the water (right) as she waits for accommodation in the dry docks.

Fruit was another major import. Manchester fruit brokers held sales at the docks, where buyers paid lower prices than at Liverpool for fruit in better condition. The Elders & Fyffe ship *Greenbrier*, displaying another connection with north-eastern enterprise via her registration in West Hartlepool, became a regular visitor, bringing in shipments of bananas. (R. Banks)

At Manchester Corporation's 'Foreign Animals Wharf' at Mode Wheel, 'lairages' were established to receive imports of live animals. Cattle leave the vessel *Manchester City* in 1898. This ship could accommodate up to 700 animals, and had extensive refrigerated space for frozen meat. The successful passage of such a large ship along the Canal did much to encourage other shipowners. (R. Banks)

The Ship Canal's connecting locks to other waterways enabled cargoes to be transhipped at Manchester and transferred into smaller craft for onward passage to more distant destinations via the narrow-boat canal network. The sailing barge *Bengal* lies alongside other barges awaiting cargo at the end of No. 7 dock.

The barge *Alice* arrives at Trafford Wharf to collect cargo. At the stern of the boat, hand on the tiller, stands the skipper, with his wife alongside. Opposite is moored the steamship *Bellasia*. Coal trucks from Bridgewater Collieries line the rails by the timber stacks. (Charles Downs)

In the early years, the Ship Canal was a novelty visited by many thousands of sightseers. As a result, a Ship Canal Passenger Steamer Company was set up to cater for those who wished to sail along the waterway. The Liverpool paddle-steamer *Toiler*, seen here passing Trafford Park as she returns unladen, was one which made regular journeys.

The passenger paddle-steamer *Ivanhoe* passes through the Barton Aqueduct and road bridge. The sealed ends of the aqueduct may be noted. Both swing-bridges were mounted on an off-centre island, the wider channel on the southern side being the one used normally by all but the smallest vessels. Passenger traffic soon failed, and investors in that service lost their money. (Grosvenor)

Small groups wishing to tour the docks were conveyed on the workmen's ferry boat, whose usual duty was to run 'every few minutes' between Trafford Wharf and other parts of the docks. The proprietor of the ferry was self-styled 'Captain' Robert Casey, who lived nearby. The postcard photographer has captioned his print, 'An interesting party viewing the sights on MSC'. All wearing top hat and tails, they must qualify as the best-dressed voyagers! (T. Pinder)

When the full length of the canal opened for traffic, the Company did not possess an adequate number of tugs to meet all requirements. Liverpool tugs were licensed to pilot large vessels along the waterway until the Company's own tugboat fleet grew to sufficient strength, whereupon the privilege was withdrawn. At the western end of the canal, a tug of the Alexandra Towing Company leads the way, while at the stern is the twin-funnelled paddle-tug *Merry Andrew*. (R. Banks)

SECTION THREE

THE TERMINAL DOCKS

In 1894 the terminal docks consisted of Nos 1 to 4 at Pomona (centre top), catering for the smaller coastal and short sea-trading vessels, and Nos 6 to 8 (centre, numbered from right to left) for the larger ocean-going ships. The docks at Pomona were on the site of the former Botanical Gardens and Zoo, and lay mainly within the Borough of Stretford, while the three large docks, within a meander of the River Irwell, were wholly within Salford. The docklands occupied 406 acres, with over 6 miles of quays. Collectively known as 'Manchester Docks', only a tiny portion of No. 1 dock lay within the Manchester boundary. At Trafford swing-bridge (right), the river had been widened to give a more generous curve on the approach to Pomona. The original course, spanned by the smaller second bridge, was filled in at a later stage. Dock No. 5 was intended to lie parallel to the main road (centre, left to right), but plans were changed and it was never completed. Instead, in 1901, the Company purchased the site of the Manchester racecourse (extreme left), in order to construct the largest dock, completed in 1905 as No. 9. Mode Wheel locks (bottom left), were the fifth and last set of locks on the 36-mile journey along the canal.

With a length of 850 feet, No. 6 dock was the shortest of the large docks. This busy scene shows vessels moored alongside each other as they wait for vacant berths. The ships include the *Kittiwake* (left), *Jabiru* (centre), with another alongside, all three being vessels of the British and Continental Steamship Company, a Liverpool-based shipping line which ran regular services to ports in Holland, Belgium and Northern France. (J.L. Brown)

Until 1905, No. 8 dock, 1,340 feet long and 250 feet wide, was the largest terminal dock. At its head was the turning basin, a large stretch of clear water to enable vessels to be swung round ready for the return journey down the canal. Viewed from across the basin, the *Langfond* (Stavanger) and another Norwegian vessel are moored (left) with the *Fanefjeld* (centre) tied up alongside a Prince Line ship. (Charles Downs)

The roof of the grain elevator on Trafford Wharf provided a good viewpoint for the photographer as he looked across the turning basin towards (left to right) docks 8, 7 and 6. In the foreground, stands of imported timber awaiting collection lie between the dockside and the railway sidings.

A 1933 aerial picture of docks 8, 7 and 6 illustrates the different dimensions of the original terminal docks. At the head of No. 8 dock (left) stands the new Dock Office, built in 1927. Across the road at the head of No. 7 dock is the Custom House, and beyond are the smoking chimneys of the Ordsall industrial area. Two vessels of the Manchester Liners' fleet are moored side by side in No. 6 dock. (Fotaire)

Pontoon docks for ship repairs at Ellesmere Port and Mode Wheel were financed by Tyneside businessmen, who also bought land to provide a graving dock (for cleaning and repairing ships). The floating docks were constructed in the north-east and towed around the coast. In the pontoon, with evident bow damage, is the *Carbineer* (1907), a steamship of the Fisher Renwick Line, which ran a regular service between Manchester and London. (Charles Downs)

In 1894 it was claimed that the turning basin, 1,388 feet across, was extensive enough to swing a ship as large as the *Great Eastern*. In 1904 the Federal Steamship Company's ship *Suffolk*, 500 feet long, was hailed as the largest ship to reach Manchester at that date. Here, the *Cardita*, of the Anglo-Saxon Oil Company, London, has been swung (paddle-tug *Eccles* at the stern), prior to entering Mode Wheel Locks.

The timber ship *Melanie Groedel* demonstrates the method of supporting deck cargo as she waits to unload by Trafford Wharf. The crew member who sent this postcard to his wife, wrote: 'This looks like a timber yard! We expect to be in Cardiff on Sunday.'

Timber continued to rank high among the imported cargoes. From the hold of the steamer *Bihac*, wooden planks are unloaded by crane and sling on Trafford Wharf in 1948. The *Bihac* (ex-*Prince Andrej*) was a 1930 Glasgow-built vessel, registered in Dubrovnik, and owned by the Yugoslav State Enterprise Line. (Kemsley Newspapers)

Viewed from the air in a south-westerly direction in 1951, Pomona Docks lie centre left, with the long frontage of Trafford Wharf and part of the main terminal docks on the top right. Factories line the water's edge on the Salford bank (right), and beyond are the tightly packed rows of

terraced housing in the Ordsall area. Trafford Park is in the top right of the scene. Near Pomona, only a few yards separated the narrow Bridgewater Canal from the Ship Canal. This is now the site of a new connecting lock intended to replace the older Hulme Lock, further upstream. (Aerofilms)

Pomona Docks provided the berthing and warehousing facilities for the shorter sea services. Fisher Renwick & Company of Newcastle began a service to London from the western end of the canal in 1892, and their distinctive three-band funnel marking became a familiar sight at Pomona from 1894. *Yeoman* (1901), was one of ten ships, all of which (apart from *Fishren*) carried military names.

No. 4 dock at Pomona, 1895. The steamship *Seal* has a narrow boat alongside as she discharges cargo. Transit sheds on the wharves handled cargoes for berths at each side. Behind the cart (left) is one of the vertical-boilered mobile steam cranes, looking elderly even at this early stage, and lacking any sort of weather protection for the operator.

In 1901, on the acquisition of the Manchester racecourse site, construction work began on what was to be the largest of the terminal docks. An embankment (right) isolated the excavation from the main canal while work was in progress. The steamship moored close to the retaining wall is the *Manchester Shipper*. The piers on which the quays were constructed became invisible once the dock was flooded. (MSCCo)

When work on the new dock was complete, water from the main canal was allowed to lap in through a small breach in the embankment. When levels were equalized, the separating wall was demolished and its remains dredged up or hammered by steam-ram into the bed. A dredger is at work (right) as a Rochdale Canal Company barge crosses the end of the new No. 9 dock. (T. Pinder)

At 2,700 feet long, No. 9 dock was over twice as long as the next largest, No. 8. At later stages, a giant grain elevator, connected to discharging berths by conveyor belts in subways beneath the quays, and additional warehouses were added. The Runciman ship, *Kentmoor* of Newcastle, is moored on the left as work proceeds on the construction of the grain elevator in about 1912. (National Series postcard)

The view from the No. 1 grain elevator shows the completed No. 2 elevator and a busy No. 9 dock full of shipping. The space between the new dock and No. 8 (right) was utilized as a marshalling yard, main line rail connections departing from the far end. The steamship *Claro* is berthed opposite. Note the stacks of imported timber on both sides of the canal.

No. 9 dock viewed from the new grain elevator in about 1920. This view shows barges moored alongside Norwegian and Danish vessels. Further along, the steamship *Turnbridge* is berthed by more timber imports. Additional warehouses were later constructed to the right of this picture. Three small children stand by the stern of one of the larger barges.

No. 1 grain elevator on Trafford Wharf was destroyed in a wartime air raid, but No. 2 survived until 1983, when it defied the first attempts at demolition. This aerial view of about 1950 shows it standing at the head of No. 9 dock. The land at the top right was later developed as a container terminal. The marshalling yard was connected to the main line railways via the New Barns junction, centre left at the bottom. (Photair)

The tanker *British Star* appeared in 1918 in First World War dazzle paint, which claimed to make it impossible for the enemy to identify ships at sea. The vessel was managed by the British Tanker Company on behalf of the Admiralty. It was able to carry 6,900 tons of petroleum in bulk, and sailed between Manchester and ports of the Persian Gulf. (Charles Downs)

The Ship Canal was an unlikely location in which to find a First World War 'mystery ship'. Shepherded by tugs, its visit to Manchester may have been for repair work in the dry dock. 'Mystery ships' were designed to appear as unarmed merchantmen, but when attacked by the enemy, flaps would open to reveal armaments with which to surprise the raider. (Charles Downs)

When the United States joined the First World War conflict, many Americans landed in Salford Docks en route to training camps inland. On 16 July 1918 the Ellerman Line ship *City of Calcutta* (built 1903, normally carrying 179 passengers), brought in 52 officers and 1,307 other ranks, returning two months later with a similar complement. (Charles Downs)

Other Ellerman ships, including *City of Marseilles* and *City of Exeter*, also brought American servicemen to Salford in 1918. Here, kitbags are loaded on to horse-drawn carts as the soldiers prepare to leave the dockside. (Charles Downs)

Heavy engineering industries in the Manchester area included several locomotive-building firms, such as Beyer Peacock, the Vulcan Foundry, and Nasmyth's of Patricroft. In 1932 fully assembled Nasmyth steam locomotives for China were being loaded on board Christen Smith's *Belpamela*, which carried heavy lift booms on the masts. (H.N. Davis)

Cayzer Irvine's Clan Line steamers were early supporters of the Manchester Ship Canal. The ships provided a service to Bombay, Colombo and Calcutta, bringing back raw cotton for the textile industry. The Company lost twenty-eight vessels in the First World War, and *Clan Robertson*, photographed at Salford in 1937, was an 8,000-ton replacement dating from 1920. (Mack & Co)

Twenty-five years separate this picture from the previous one. In 1963 the Brocklebank Line's *Mahout*, on the Manchester–Calcutta service, lay in No. 8 dock with one of the floating pneumatic grain elevators alongside. The floating elevators sucked out grain from the hold and discharged it into lighters moored alongside. The edge of the Dock Office appears on the extreme right. (Elsam, Mann & Cooper)

The Danish short-sea vessel *Procyon* (built in 1938), slices through the placid waters of the canal as she passes the dry docks on her way to Pomona in 1966. The graving docks are occupied by two Manchester Liners. The stern of the *Manchester Engineer* (ex-*Cairngowan* of 1952) may be seen on the left. (E. Gray)

Pomona Docks, catering as they did for the tramps of the coastal trade, tended to acquire a run-down appearance to match. The smart ocean-going liners did not deign to appear here. On this occasion in the early 1930s, however, even the elderly sand and gravel boat *Assurance* is dressed overall as the tug *Ralph Brocklebank*, with an inspection party on board, tours the docks. (MSCCo)

Coast Lines Limited was an amalgamation of several coastal shipping companies, which operated between most British ports using an extensive fleet of steam and motor vessels, nearly all of which carried a name with the suffix 'Coast'. The motor ship *Denbigh Coast* of 1937 was photographed here in Pomona Docks, her cargo being unloaded by an ancient steam crane. (Bert Wilson)

Barges from the Bridgewater Canal were able to transfer to the Ship Canal via Hulme Lock, nearer to Manchester's city centre. Lighters also served premises, such as the Salford refuse disposal yard, which lay further upstream, beyond the usual limit of navigation. The tug passing the entrance to Pomona Docks in 1954 had collected this laden lighter from some point upriver. (E. Gray)

In Pomona Docks in August 1954 was the *Multistone*, an elderly steamship dating from 1910, owned by Robert Gardner of Lancaster. The grab crane is in use as she discharges china clay from Cornwall. In the right foreground is the stern of the paddle-tug *Rixton*, at that date withdrawn from service and laid up pending sale. (E. Gray)

Changes in the shipping world of the 1970s are typified by the container ship *Manchester Falcon*, berthed in No. 9 dock in 1975 opposite the Harrison liner *Explorer*. Escalating shore costs and labour problems had accelerated the transition to containerization, and the increasing use of the new methods was to render the traditional ship design redundant. (Norman Edwards Associates)

Section Four

Trafford Park

Trafford Park, ancestral home of the de Trafford family, occupied a unique position, almost an island site, between the River Irwell and the Bridgewater Canal. The latter had been dug along the Park's southern boundary in 1761. In the early 1880s Sir Humphrey de Trafford had proved an implacable opponent of proposals to convert the river into a ship canal, for he did not wish to have his domestic peace disturbed. After his death in 1886 his successor (eldest son, Humphrey Francis) proved more amenable to negotiation, and agreed to sell a portion of the estate at the Manchester end of the Park. This was to become Trafford Wharf and, as a result, the plans for the docks were recast. Indeed, it appeared that the new Sir Humphrey was willing to dispose of the whole estate, and Manchester Corporation considered its purchase for use as a public park. However, in 1896 Trafford Park was bought by speculator E.T. Hooley, with the intention of converting it into an industrial estate, offering the advantage of easy access to deep-water berths for ocean-going ships. This 1930s view of the Manchester end of the park illustrates how the land between the Bridgewater Canal (right) and docks became crammed with factories. (Altigraph)

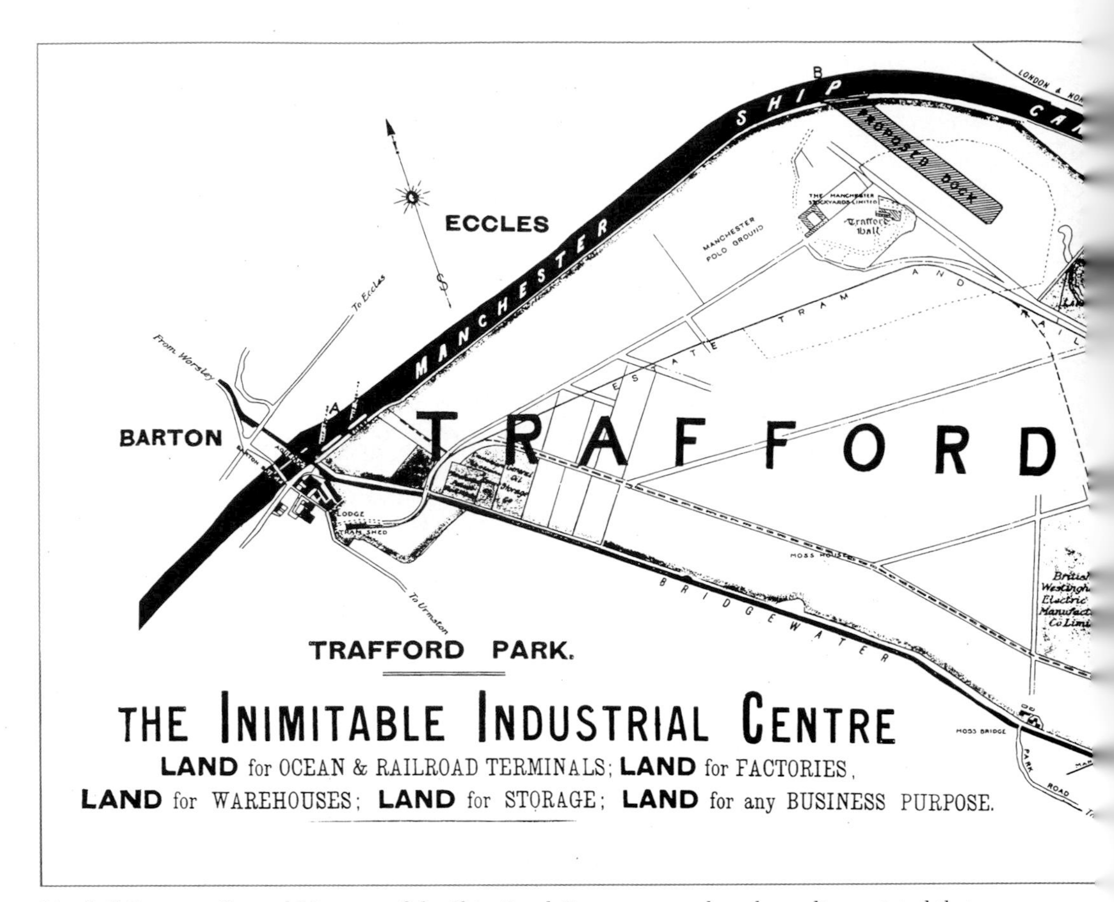

Marshall Stevens, General Manager of the Ship Canal Company, may have been disappointed that the new waterway was not an instant success, and consequently he accepted the challenge of a new post as Manager of the Trafford Park Estates. His task was to attract industrialists to the Park and persuade them to use the canal. Publicity claimed that manufacturers who located their industries in Trafford Park could make immense savings in the cost of transporting raw materials

The location of the oil tanks on the Trafford Park bank at Mode Wheel demonstrates the advantage of direct transfer between ship and installation, with no delay or intermediate costs of transport involved. Tanker *Varo* is discharging her cargo, while on the opposite bank a vessel is moored at the wharf of the Union Cold Storage Company.

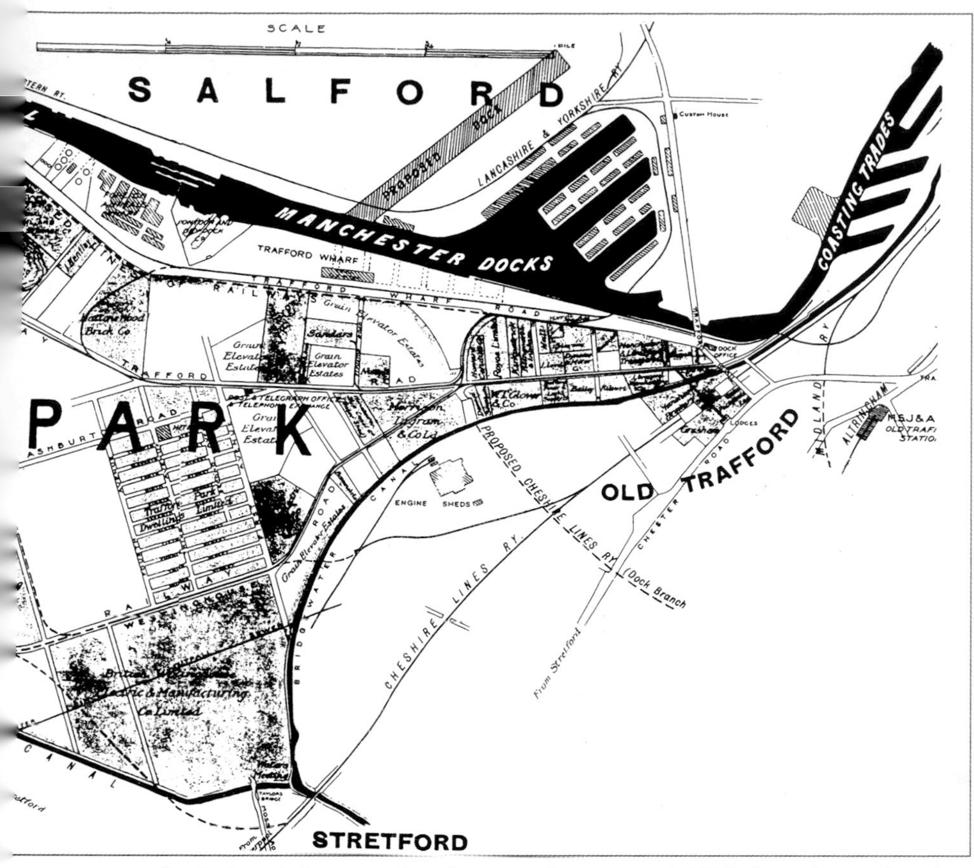

or finished goods between ship, factory and consumer. A 1903 plan indicates that plots nearest to Manchester were occupied first, while the western end of the park remained rural in character, offering leisure pursuits such as golf and polo, with fishing and boating on Trafford Lake. The proposed dock marked on the plan by Trafford Hall was never built, but that on the Salford side became the large No. 9 dock in 1905.

Thomas Hedley's soapworks (a subsidiary of Procter & Gamble) took land close to Trafford Hall with a frontage on the canal, and installed an overhead pipeline to link with steamers berthed on the opposite bank. The Norwegian ship *Mantilla* discharges palm oil in the 1930s.

The Manchester end of Trafford Park lies on the left half of this 1950s aerial view of the terminal docks. The Park was well served with rail connections to all factories. The railway swing-bridge across the canal,

which linked the Trafford Park lines with the other terminal docks, is bottom left. Mode Wheel Locks are centre top. (Airviews)

The Corn Products Company of Trafford Park (now Cerestar, originally Nicholls Nagle) constructed its own wharf on the canal (top left). The Thomas Hedley soapworks is centre right. Trafford Hall (bottom right), once a stately home in the centre of a magnificent deer park, survived for some years as a hotel and golf clubhouse. Though latterly unoccupied, it withstood the encroachment of the factories until after the Second World War. Traces of the former paths and gardens can be seen in this 1930s view.

The premises of the Corn Products Company (centre left) were rebuilt, the plant enlarged, and the wharf modernized to receive supplies of corn in bulk. The Thomas Hedley soapworks, now Procter & Gamble, lie centre right. The Centenary Bridge, opened in 1994, now crosses the canal at a point between the two factories, linking Trafford Park with Eccles. Again, traces of the former Hall gardens can be seen.

Sir Humphrey de Trafford had insisted that the canal bank on the Trafford Park side should have a slope of 22.5 degrees, at the top of which the Canal Company must build a wall, not less than 9 feet high, to protect his estate. A length of this wall may be seen on the unoccupied land on the near bank in this 1952 view. A Strick Line vessel is passing the Dutch tanker *Juno*, moored on the Weaste side. The rail connection to the former LNWR main line is at the top right. (Airviews)

The same location as above, but viewed from the Weaste side of the canal looking south towards Trafford Park, has the Swedish tanker *Carina* moored at the Berry Wiggins oil wharf. The Thomas Hedley factory is to the right. (R.L. Wilson)

The private wharf of the Cerestar factory in Trafford Park has equipment for sucking out bulk grain from the holds of vessels and discharging it direct into the plant, a procedure which is exactly what the pioneers of the industrial estate envisaged in 1896. The *Carchester*, a bulk grain carrier, is seen berthed at the wharf in 1969. At the same wharf today the direct discharge process continues, for Cerestar still imports its supplies in bulk via the canal, receiving two or three shiploads per week. (MSCCo)

SECTION FIVE

THE WESTERN PORTS & INTERMEDIATE WHARVES

The 'Port of Manchester' includes the whole length of the Ship Canal, from its entrance at Eastham to the terminal docks 36 miles away. The canal may thus be considered an elongated harbour. At its western end the docks at Ellesmere Port and Runcorn, both formerly with free access to the tidal Mersey Estuary, were already established as busy ports long before the Ship Canal was constructed. Ellesmere Port was the terminus of the Shropshire Union Canal, and consequently an important transhipment point, while Runcorn performed the same services for cargoes to and from both the Bridgewater Canal and the Mersey & Irwell Company's Latchford Canal. Vessels from the Weaver Navigation also enjoyed unobstructed entry to the Mersey. Consequently, uninterrupted traffic to these points was maintained for as long as possible during the Ship Canal construction period, and the first completed sections of the canal were brought into use at the western end in 1891. Sailing vessels crowd the old wharves at Runcorn in a view from this early period. The *Tregunnel* of Padstow is moored left. When the full length of the canal opened, sailing ships with masts too high to pass under the fixed bridges, transferred their cargoes to lighters at Runcorn.

Ellesmere Port docks were leased to the Ship Canal Company, and the Company was responsible for the work of loading and discharging vessels. The German ship *Carl Cords* discharges timber, while the *Artigas* of Philadelphia unloads grain. The legs of the dockside cranes straddled the railway tracks. (R.M. Morris)

A later view of the Ellesmere Port wharf shows the Canal Company's electric cranes, which replaced the older equipment, loading crates aboard the Swedish vessel *Ring* of Stockholm. On the wall of the warehouse, faded lettering advertises the services of the Shropshire Union Canal Company. (Elsam, Mann & Cooper)

Two miles along the canal, between Eastham and Ellesmere Port, the Bowater's UK Pulp and Paper Company established a wharf alongside its mills to unload wood pulp and minerals from ocean-going vessels. The Bowater ships all carried family names. The *Elizabeth Bowater* was built in 1958 and displays the Company's insignia on the bow. (S.H. Thomasson)

Imports of crude petroleum featured increasingly among the Canal's regular traffic, so in 1922 a specialist oil dock was constructed in an isolated area on the estuary side of the canal at Stanlow. Exports of petroleum products grew, and a second and larger dock was built alongside the first in 1933. In this 1947 view two large tankers occupy the second dock, with two coastal tankers in the original one. (Airviews)

The Eagle Oil Company's tanker *San Gaspar* (new in 1921), lies alongside the Norwegian ship *Spinanger* in Stanlow's No. 2 oil dock in the mid-1930s. Oil was pumped from the ships to the refineries though pipes laid beneath the canal. (W.R. Roe)

The short-sea vessel *River Avoca* (ex-*Stevonia* of 1948) bought by R.V.T. Hall of Eire in 1962, passes the Stanlow oil docks on its outward journey along the canal. The tidal flats of the River Mersey lie beyond the oil docks, which were sited on the estuary side for safety reasons.

From Stanlow, oil berths proliferate for over a mile upstream to cater for the consumption of the ever-expanding refineries of the Wirral. Direct pipelines link the wharves with the storage tanks of all major oil companies. The coastal tanker *Shell Director*, built in 1972 for Shell UK Limited, is at the Ince oil berth in 1984. (E. Gray)

In 1954 the Canal Company opened a new four-berth oil dock at Eastham, with direct pipeline links to the Stanlow refineries. The Queen Elizabeth II Dock has a separate entrance lock, 807 feet long and 100 feet wide, and is able to admit tankers too large to enter the canal proper. The wide beam of the tanker moored left in the new dock may be compared with the narrower vessel about to leave the canal via the original locks in the foreground. (Stewart Bale)

The Weston Point Docks of the Weaver Navigation connect with the Ship Canal on the bend approaching Runcorn. A signboard for the smaller waterway proclaimed: 'Shortest Route To The Potteries. Quick Despatch. Bunker Coal Supplies.' These small docks lie behind Alfred Holt's Blue Funnel Line steamer *Neleus* (1911), which is berthed at the Imperial Chemical Industries' salt works wharf.

At Runcorn the old docks had been acquired along with the Bridgewater Canal, and so were administered by the Company. Because of the restricted entrance width, and the limited depth of water, they catered mainly for smaller vessels. A steam crane with tipper bucket, its operator protected by corrugated iron sheets, is unloading lump china stone from the Hughes Holden ship *Moelfre Rose* at Flint Wharf in the 1930s. The dock's entrance from the Ship Canal is mid-right.

Runcorn Docks enjoyed a busy period in the 1970s when the motorway network was extended into north Cheshire. An expressway provided easy access, and a widened dock entrance facilitated its use by vessels up to 350 feet long drawing not more than 17.5 feet of water. Vessels in the crowded dock include the *Saint Bridget* (right) and the Dutch vessel *Plancius*. In the distance the graceful arch of the 1961 road bridge curves behind the 1868 railway viaduct. The latter, approached by a sharp bend, lay on the narrowest portion of the canal. (MSCCo)

Runcorn bridges viewed from the inland side, October 1970. The narrow embankment separating the canal from the tidal estuary is to the right. A Manchester Liners' container ship, the then new *Manchester Merit*, makes her way along the canal, unaided by tugs. Runcorn Old Quay repair yard is to the left, while on the right is Old Quay Lock, one of three disused locks along this stretch which once allowed small boats direct access to the Mersey. (Norman Edwards Associates)

In the early years the largest refuelling point on the canal was the Partington coaling basin, where equipment allowed the contents of railway wagons to be tipped directly into ships' bunkers. Empty wagons were hoisted to a line above, from where they returned to the sidings by gravity. The coaster *Bellrock* receives coal from a Bickershaw Collieries wagon. (R. Banks)

The Strick Line ship *Baharistan* (new in 1928) passes the Partington coal tips as she traverses the canal in 1951. In the days when most ships were powered by steam, there were seven coaling stages and over nineteen miles of railway sidings at Partington, on which locomotives and steam cranes were permanently employed in replenishing supplies received via the main line railways. (H. Cullen)

A little further inland from Partington coaling basin was the Lancashire Steel Corporation's wharf at Irlam. The *Pencarrow* of Falmouth was discharging iron ore in 1949, as the tug *Cadishead* passed with the outward-bound steamship *Mercy Warren*. The latter was a 1943 Portland-registered vessel of the United States War Shipping Administration. In the distance is the Irlam railway viaduct and locks. (MSCCo)

At Davyhulme in 1889, Manchester Corporation purchased land from Sir Humphrey de Trafford on which to site a giant sewage farm. When the canal opened, a specialist wharf was constructed, where sludge was transferred to a vessel which conveyed it along the canal for dumping out at sea in the far reaches of Liverpool Bay. The Corporation's special steamer for this task was the *Joseph Thompson*.

The sludge vessel *Joseph Thompson* was built in 1897 by Vickers of Barrow. This deck view in Irlam Locks shows that, despite her task, which she carried out several times per week, she was kept remarkably clean. Note the three officers on the open bridge and the deckhand ready with the mooring rope.

Salford Corporation's 1876 sewage plant at Weaste treated material which had formerly been discharged direct into the River Irwell. After the opening of the canal Salford, too, owned a vessel, the *Salford* (1895), which carried waste matter to be dumped out at sea. In 1928 the original vessel was replaced by the *Salford City*, here decorated on the occasion of an 'inspection' – that is a trip down the canal for local councillors.

Waste material continued to be carried down the canal for disposal at sea until 1987, when the construction of a pipeline from the treatment plants was completed. The *Gilbert J. Fowler* was one of two North West Water Authority sludge boats (the other was the *Percy Dawson*) employed on this regular task in 1984. (E. Gray)

An unusual visitor to the canal in 1952 was this whaling factory ship *Polar Chief* (ex-*Anglo-Norse*), owned by the Falklands Whaling Company Limited and registered at Leith. She was discharging whale oil while berthed at Little Bolton, Weaste, opposite the Thomas Hedley soapworks. (R.L. Wilson)

The *River Andoni*, a 1979 vessel of the Nigerian National Shipping Company, Lagos, became an object of curiosity while moored for over two years, from 1994 to 1995, at a wharf in Ellesmere Port. The vessel had been impounded because of non-payment of debts by her owners. Several of the Nigerian crew, unable to reach home, remained on the ship and were befriended by local residents. (E. Gray)

LOCKS, BRIDGES & SPECIAL FEATURES

Along the projected route of the Ship Canal lay several obstacles. Manchester is 60 feet above sea level, so sets of locks were required. Established road crossings had to be accommodated, and there were four places where railway lines crossed the proposed course on the level. (At a fifth location, the LNWR's 1868 Runcorn viaduct already had sufficient height to clear the masts of any ships passing beneath.) There was also the difficulty of replacing Brindley's 1761 Bridgewater Canal aqueduct across the river at Barton. Approved plans included an entrance lock to seal off the canal from the tidal estuary at Eastham, and four other locks each to have a rise of approximately 15 feet. At major road crossings, swing-bridges were to be provided. For the railways, swing-bridges were rejected for safety reasons, and tunnels on the grounds of expense. Instead, long embankments were proposed for the approaches to the canal to lift the rail track to a minimum of 75 feet above water level. To replace the Bridgewater aqueduct, Leader Williams first suggested two boat lifts, one on each side of the canal, similar to that already in use at Anderton, linking the Trent & Mersey Canal to the Weaver Navigation. This 1951 view of Eastham shows, top to bottom: the small barge lock, the medium-sized lock, a steamer in the large lock and (nearest the camera) the approach channel to the sluices. The outward gates of the middle lock are open to the estuary. Note the double set of gates (one set folded back into the recesses) fitted at the outward ends as a safety measure. (Valentine)

Eastham entrance locks were normally operated from four hours before to four hours after high water. On what was known as a 'levelling tide', Mersey water would flow into the Ship Canal, but the usual function of the lock gates was to maintain the standard depth of water in the canal. A vessel led by Liverpool tug *Langarth* approaches the entrance to the large lock on a high tide. The railway-type signal indicates which lock should be used. Note again the two sets of gates at the outer end.

Once inside the large lock, the water level would be adjusted if necessary to match that in the canal, and the ship would then proceed. Certificated pilots could be engaged by ships' masters unfamiliar with the canal. Steam tug *Mercia* and the paddle-tugs *Barton* and *Eccles* await the call to duty. The approach channel to the sluice gates proved a useful mooring point for tugs.

The France Fenwick Company motor vessel *Sherwood* (new in 1958) is towed from the large lock by the tug *Sovereign*, while another vessel waits to enter. Each time a lock is used, a volume of water is lost to the lower level, so at Eastham the three chambers of different sizes were designed to minimize that loss. At each of the other four locks there were only two chambers.

A major task which, fortunately, had to be carried out only on rare occasions, was the replacement of lock gates. In 1939 a 250-ton crane, accompanied by a Ship Canal tug, was hoisting a new gate into position. Liverpool tugs normally shepherded vessels in the approach channel to the canal, Company tugs taking over thereafter. (MSCCo)

The Latchford locks are twenty-one miles from Eastham. Immediately before the locks, the canal passes beneath a railway viaduct, which, like those at Acton Grange, Cadishead, and Irlam, was constructed in 1893 close to the original route. For testing purposes, the new viaducts were used for several months by goods trains only. During this trial period, the original lines were left in place on the level for use by passenger trains, and excavation of the canal at these points was consequently delayed. This 1954 aerial view illustrates why the canal was referred to as 'The Big Ditch'. In the foreground, an oil tanker makes its way towards the sea, while the small lock chamber is open to the upper level, ready to receive another outward-bound ship.

Railway tracks over the canal were raised by the construction of long embankments, beginning over 1¼ miles away so as to offer a steady upward gradient. The steamer *Ionian Trader* (a Panamanian-registered 'liberty ship', headed by tug *Firefly*) demonstrates the limited clearance as she passes beneath the Latchford viaduct in 1954. At Eastham a de-masting crane removed, if necessary, the tops of non-telescopic masts or funnels from vessels proceeding beyond Runcorn.

Latchford locks, viewed from the railway viaduct in 1958. The steamer *Sandsend* is about to leave the large chamber, while the inward-bound *Baskerville* is held by tugs until the lock is clear. The tug in the foreground is the *MSC Onset*. The extra width of the canal on the approach to the locks allowed space for a useful lay-by, or mooring point.

At Warburton was the second of two cantilever bridges which carried minor roads over the canal. The other, at Latchford, was similar. Both had steelwork weighing 783 tons, and a centre span of 206 feet, with a clearance above water level of 75 feet. Warburton Bridge proved a good viewpoint as tug *Cornbrook* led the American steamer *Charles Lykes* through the Cheshire countryside in 1957.

Irlam Locks lie some seven miles further upstream from Latchford, and have a similar construction. The large chamber measures 600 by 65 feet, and the smaller 350 by 45 feet, with sluices to the side (see page 16). The sand and gravel boat *P.M.Cooper* approaches the smaller lock (right), while the Monroe Brothers' boat *Kylebank* (built 1925) is moored in the lay-by. (Lilywhite Ltd.)

The 65-feet-wide lock chambers had been thought generously proportioned in 1894, but as ships' beams increased, clearances became less, and some Manchester Liners' vessels in the 1970s were designed to fit the limited dimensions of the locks. The *Manchester Reward*, outward-bound with containers for Montreal in 1979, demonstrates the tight fit as she edges into the large lock at Irlam. The tug *Scimitar* is in attendance. (Norman Edwards Associates)

Barton Locks, 1949. The fourth locks lie only a short distance on from Irlam. The Furness Withy liner *Pacific Enterprise* is drawn out of the large chamber (left) by the bow tug. Her stern tug, having descended in the small lock, waits to be reattached. Two other tugs hold the inward-bound Harrison liner *Sculptor* as she waits to enter the vacated chamber. (Airviews)

The fifth and final lock on the canal is that at Mode Wheel, thirty-four miles from Eastham, and the gateway to the former terminal docks. In 1933 this vessel approaching Mode Wheel passed the British Empire Steam Navigation Company's ship *Gambia River* (left) berthed at the Cold Storage Company's wharf. The Trafford Park bank is to the right, with the lock entrances directly ahead. (Entwistle Thorpe)

Looking downstream from the end of Mode Wheel Locks, in the opposite direction from the previous picture. An outward-bound steamer, which has just descended in the large lock, passes the frozen meat stores (right) as the tugs of the Liverpool steamship *Cento* prepare to position her to enter the empty chamber. (J.L. Brown)

Mode Wheel Lock, April 1935. The Liverpool barges *Arabia* and *Siberia* occupy part of the small lock. Looking upstream, the large lock chamber is to the left, with the lock-keepers' cabin between. The '200 Feet Stop' sign, left, was to indicate mooring positions for shorter boats. The intermediate lock gates, an aid to conserving water supplies, and also an additional safety measure, are tucked into the quaysides by the barges. (A.H. Clarke)

Mode Wheel, looking towards Manchester, about 1950. The differing sizes of the lock chambers are clearly defined, and, further on, is the expanse of water in the turning basin. The long No. 9 dock with the grain elevator at its head, is top left. The tanker moored below is at the Mode Wheel oil wharf on the Trafford Park bank. (Airviews)

In the Victorian era the twentieth-century growth in road traffic was not envisaged, and the Company felt confident in providing swing-bridges at the seven points where major roads crossed the route of the proposed canal. Barton Road swing-bridge (seen here with the swing-aqueduct behind) was a typical example. In the early days hydraulic operation was adopted.

The Northwich Road swing-bridge at Stockton Heath, Warrington, seen here in 1930, is closed to road traffic as it swings to allow the passage of the Danish ship *Ragnhils*. Staff closed barriers across the road before swinging the bridge. In the early days, signals to ships were by means of baskets hoisted on poles fitted to the upper girder work. (Valentine)

Trafford Road, Salford, boasted the largest, the furthest inland, and the busiest bridge for road traffic, though only those ships going to Pomona Docks passed through on the canal. In 1905 the Canal Engineer adapted the bridge to accommodate Salford's electric tramway route to Trafford Park, ensuring the alignment of track and overhead wires each time the bridge swung. The Coast Lines vessel *Moray Coast* (1905) sails through in the 1920s. (J.L. Brown)

Crossing points not deemed important enough to warrant the construction of a bridge were supplied with a ferry, sometimes, as at Bob's Ferry, Cadishead, merely a rowing boat. The *MSC No.1 Horse Ferry* was at Irlam, and in this 1920s view is transporting a motor car. At a later stage motor launches were provided, until they proved uneconomic. (H. Grundy)

Engineer Leader Williams changed his plans for conveying Bridgewater Canal barges across the Ship Canal. He abandoned the idea of two boat lifts, whose use would have been time-consuming, in favour of a swing-aqueduct. This was designed as a sealed tank which, when the ends of the Bridgewater Canal were also sealed, could be swung while remaining full of water. The *Manchester Division* passes through the Barton bridges in 1951. (Airviews)

A 1963 view of the same location, but from the opposite bank, shows the sealed ends of the Bridgewater Canal and road traffic queueing for the road bridge to reopen. The overgrown remains of the old embankment leading to the former stone aqueduct may be discerned in the centre foreground, between the narrow canal and the road. The former Barton power station (top right) received much of its coal via Bridgewater Canal barges. (Airviews)

Barton, looking towards Manchester, in 1967. Road bridge and aqueduct have swung to allow the passage of an inward-bound ship. A tanker lies at the Barton oil berth and, further along on the opposite bank, vessels are moored at the Irwell Park Wharf, which specialized in the discharge of dry bulk cargoes, such as sulphur or ore, and from where cargoes of scrap metal were exported. On the Trafford Park bank, by the bend in the canal, are the premises of the Corn Products Company and Proctor & Gamble's soapworks. (Airviews)

In 1893 the Ship Canal Company sold a piece of land near Mode Wheel for a Tyneside syndicate to establish the Manchester Dry Docks Company. Three graving docks were constructed, and a floating pontoon dock was available nearby. In this 1920s picture all four are occupied by ships under repair or inspection. (Imperial)

Another view of the dry docks, from a different angle, shows the nearest dock unoccupied. A Coast Lines vessel sits in the floating dock. The sluices of Mode Wheel lock controlled the level in the upper basin, surplus water from the River Irwell being allowed to flow over. The large lock chamber is occupied by a steam-tug and a collection of barges. (Photair)

The dry docks were constructed to the same width (65 feet) as the locks, the largest dock being 535 feet long. As the water was drained from the graving dock, the ship's keel came to rest on a central line of 4-feet high blocks, the ship's sides being supported by stout timbers. The water level has fallen far enough to reveal the propellers of the Manchester Liners' chartered vessel *Frontier* (later *Manchester Frontier*) in 1972. (Norman Edwards Associates)

Some years after the opening of the Ship Canal a transporter bridge was constructed across the Mersey Estuary to link the towns of Runcorn and Widnes. This was not a Canal Company venture, but it remained a prominent feature of the waterway for over fifty years. Opened in 1905, it was positioned on the inland side of the railway viaduct. High towers on each bank supported a high-level girder span fitted with rails on which ran an electrically powered trolley. Suspended from the trolley was a travelling platform, or cradle, capable of carrying people, animals, goods and vehicles. On the Runcorn side, the transporter passed at low level over the Ship Canal, and the operator had to pay strict attention to avoid shipping movements on the waterway. This early picture shows the travelling platform arriving. Among the passengers is a horse and cart. The pavilion-like structure on the left was a shelter, or sitting room, for passengers during the crossing. The controller's cabin is the ornate structure perched centrally on the arched steelwork. Only a flimsy chain protects the open edge of the platform. (W. Hall)

The transporter bridge at the Runcorn bank, with tug *Archer* and the outward-bound Brocklebank steamship *Malabar*, demonstrating why the controller of the travelling platform would have to wait for the vessels to clear the bridge before he started his crossing. By the 1950s the transporter bridge was proving incapable of meeting demand, and it was replaced by a new road bridge in 1961. (Lilywhite Ltd)

SECTION SEVEN

TUGS, DREDGERS & MISCELLANEOUS VESSELS

The Manchester Ship Canal Company maintained a fleet of tugs which could be hired to assist ships in their journey along the waterway. Regulations restricted speed in the canal, and larger vessels were not able to attain sufficient momentum to give steering way. Pilotage was not compulsory, but qualified pilots and helmsmen could be hired if the master, or mate, did not possess an appropriate pilotage certificate, or was unfamiliar with the complexities of the canal. A pilot, if employed, assumed responsibility for the passage of the vessel along the canal, and in the early days would communicate with accompanying tugs by means of signals on the ship's whistle. The Company's first tugs were those acquired in 1887 when the Bridgewater Navigation was purchased. The Bridgewater undertaking had nine steam-powered paddle-tugs (dating from 1857 to 1877) for towing strings of barges from its canal terminus at Runcorn across the Mersey Estuary to Liverpool. The 1857 *Earl of Ellesmere*, seen here in the Mersey, was one of the original Bridgewater vessels which formed the nucleus of the Company's tugboat fleet. This vessel, along with three other ex-Bridgewater tugs of its type, remained in use until broken up in the late 1920s. The characteristic funnel markings of two white bands on black were a Bridgewater device, adopted and continued by the Ship Canal Company.

The Bridgewater Company owned twenty-seven smaller tugs dating from the 1870s, all named after places in Lancashire. Originally confined to the narrow canal, they continued to work for the Ship Canal Company's Bridgewater Department, and were frequently to be seen in the main terminal docks when delivering or collecting barges carrying transhipped cargoes. The tug *Walton* is in dry dock at Runcorn.

A rare picture of the *Florida*, an oddity in the tugboat fleet. Built at Leith in 1887, she had been used by the contractor in the early stages of construction of the canal, and was acquired in 1891 after T.A. Walker's death, when the Canal Company assumed responsibility for the completion of the work. She was retained thereafter for general duties until broken up in 1913.

In 1895, to supplement the ex-Bridgewater tugboat fleet, the Canal Company added two more vessels and also acquired four small screw-propelled tugs for barge-towing and general duties. The small tugs, of only 33 tons, carried ladies' names – *Eva*, *Hilda*, *Minnie,* and *Gwen*. *Eva* was photographed in the small dry dock at Runcorn in 1954. She was broken up five years later. (E. Gray)

With a tonnage of 103, tugs *Eastham* and *Partington* were supplied by the Southampton firm of J.C. Fay & Co. in 1899. From 1896 until the 1920s, the Company followed the practice of naming its new tugs after places along the canal. The *Partington*, seen here leaving Mode Wheel Locks, remained in service until 1946.

The locks, sharp turns, and narrow portions of the Ship Canal presented problems for the towage of large vessels. The old Bridgewater tugs could not drive their port and starboard paddle wheels independently, but six new tugs acquired between 1903 and 1907 did not suffer this disadvantage, and thus proved extremely suitable for stern towing. The *Barton* of 1903 served until 1950.

The *Acton Grange*, built in 1907, was another of the six paddle-tugs designed and purchased specifically for stern-towing duties. She is here seen guiding the stern of the steamship *Leka* as she leaves Mode Wheel Locks and arrives in the terminal docks in about 1910.

Dredging was necessary to remove accumulated silt and debris and maintain a sufficient depth of water for the deep-draughted vessels using the canal and its approach channels. The old *Number One* hopper, built in 1874, used until 1915, and described in its Ship Canal registration of 1894 as 'Mud Barge', was never given the dignity of a proper name. She carries out her task with crane and grab in the approach channel to Eastham Locks. (MSCCo)

The dredger *Irk* (named after the river) was one of several which operated on the endless 'chain of buckets' principle, dragging up material from the canal bed, and dumping it into barges alongside for disposal elsewhere. Allied to the dredging activities were the responsibilities of the lock staff in setting the sluice gates to maintain the correct water level. (H. Darbyshire)

The Canal Company had its own police force and fire brigade. From the early years a special vessel carrying Merryweather fire-fighting equipment, the *Firefly*, was maintained in a permanent state of readiness. Her pumps were capable of throwing out 4,000 gallons of water per minute. A warehouse fire in 1911 saw her in action in No. 6 dock. (Charles Downs)

When not in use as a fireboat, the *Firefly* sometimes doubled as the canal tender for visiting parties at the docks. The brass nozzles of the hose-pipes sprout from among the heads of this well-dressed all-male party as the *Firefly* approaches Trafford Wharf. The swing railway bridge is in the background. (Charles Downs)

The original fire tender was replaced by a second *Firefly* in 1935. Built in the Henry Robb yard at Leith, she, too, had Merryweather fire-fighting equipment, but had been designed more as a dual-purpose vessel, and so could undertake general duties. Here, she exercises her hoses while moored at Trafford Wharf. The *Manchester Commerce* lies astern. (Fox Photos)

The second *Firefly* performed invaluable fire-fighting, salvage, and bomb disposal tasks during air raids on the docks in 1940–1. She remained in service until broken up in 1965. Acting as a general service tug, she is seen here towing the Harrison liner *Craftsman* along the canal near Latchford in 1954. (E. Gray)

The Canal Company's dockside equipment was supplemented by self-propelled floating cranes, which could lift and transport heavy, awkward or bulky items for hoisting over the off-sides of ships. In No. 8 dock in 1950, a 60-ton crane lifts a railway coach for export to Portugal. (Bert Wilson)

The floating cranes usually made relatively short journeys around the docks. On longer excursions, as they were rather slow and unwieldy, they were shepherded by tugs. This 250-ton crane carries a lock gate, and has three tugs in attendance as she passes Old Quay, Runcorn. The two lead tugs are *Cadishead* and *Mount Manisty*, while one of the paddle-tugs is at the stern. (MSCCo)

The tug *Ralph Brocklebank* was built in 1903 to tow barges for the Shropshire Union Canal Company from their docks at Ellesmere Port across the Mersey to Liverpool. In 1922 she was acquired (along with two others) by the Ship Canal Company, and for some years acted as tender for visiting dignitaries. King Fuad of Egypt was on board during his tour of Lancashire in the early 1930s. The Prince Line's *Syrian Prince* has also put out the flags for the occasion. (Sport & General)

Barton, 1955. In 1936 the *Ralph Brocklebank* was completely refitted and, as hitherto the Company lacked any form of tribute to the canal's main originator and first chairman, she was renamed *Daniel Adamson*. The observation deck was later enclosed, and she is now preserved at the Boat Museum, Ellesmere Port. (E. Gray)

In 1926 two tugs were purchased from a Plymouth operator. Both had been built for the Admiralty in 1917 as *HS30* and *HS31*. On joining the Ship Canal fleet, they were named *Mount Manisty* and *Cadishead*. Both served until 1961. The *Mount Manisty* is seen from above as she passes beneath Warburton Bridge in 1954. (E. Gray)

In 1938 the Company introduced an alphabetical system of naming new additions to the tugboat fleet. The first was *Arrow*, one of four single-screw tugs of 144 tons and 750 horsepower built by Henry Robb. The eight crew members who delivered *Arrow* from Leith via Loch Ness and the Caledonian Canal claimed to have seen the Loch Ness monster during their voyage. The monster was, they said, longer than the boat and created a great wash as it swam away. (Pictorial Press)

The *Bison*, delivered in 1939, was another Henry Robb product similar to *Arrow*, which arrived in time for use in the 1939 emergency. Technically, her name was *MSC Bison*, for, since 1927, the Company had added 'MSC' as a prefix to the names of all new tugs. She remained in service until 1966.

MSC Quarry, seen here at the bow of the 1954 Anchor Line ship *Hazelmoor* (with *MSC Undine* at the stern), was a 1951 development of a new breed of powerful twin-screw oil-engined tugs, built once again by Henry Robb and fitted with Crossley engines. Their arrival enabled the Company to withdraw some of the older steam-tugs, including all six paddle-tugs.

Tugs of a new hydroconic design were heralded by the arrival of *MSC Sabre* (147 tons) in 1956. Built by the P.K. Harris yard at Appledore, the spacious wheelhouse offered a clear all-round view, the traditional funnel having given way to a tripod mast. In No. 9 dock in 1970, *Sabre* approaches the Ellerman liner *City of Lancaster*.

Between 1950 and 1960 the Ship Canal Company purchased thirteen small tugs for assorted special duties, including attachment to the dredging department. The last of the line was *MSC Dawn* (37 tons), built by Pimblott of Northwich with a Gardner engine, and seen here passing the oil docks at Stanlow in 1984.

MSC Ulex (along with *MSC Undine*) arrived in 1965 from the R. Dunston yard at Hessle. They were fitted with Ruston & Hornsby machinery. After their delivery no more new tugs were ordered for almost a decade. The legs of the mast acted as exhausts for the engines. *Ulex* is seen in 1971 approaching Eastham.

The last new vessels ordered by the Company were four V-class tugs built by J.W. Cook & Co. of Wivenhoe, in 1974–6. *MSC Victory* churns up the water in Runcorn Docks in 1976, as a tanker passes in the main canal beyond the dock entrance. The vessel moored at the quay (left) is the *Vega De Danubio*. Since 1989, the remaining tugs have been operated by Carmet Tugs Limited. (MSCCo)

Any collision or sinking on the canal which might result in a blockage was potentially very serious for the Company, so emergency procedures were drawn up to deal with any possible accident or emergency. A priority would be to remove the obstruction and/or repair the damage as rapidly as possible. Fortunately, such incidents were rare. The steamer *Onyx* had suffered bow damage in a collision with the British & Continental Steamship Company's *Dotterel*, but was able to be tied up in Pomona Docks pending admission to the dry docks for repairs. (A. Hirst)

The Norwegian timber ship *Borgfred* (ex-*Winroth*, built in 1920 by Swan Hunter) created what was potentially a more serious threat when, proceeding inward-bound without the assistance of tugs, she suffered a loss of steering power and became jammed across the canal near Barton in September 1954. Fortunately, tugs managed to free her without serious delay. (Hallawell Photos)

On 21 March 1961 the sand-hopper *Mary P. Cooper* (right) was in collision with the J.S. Monks steamer *Foamville*. The former vessel sank to the canal bed. Small and medium-sized boats managed to squeeze past the obstruction, but larger ships were trapped in the upper reaches of the canal for three weeks.

The salvage vessel *Dispenser* was summoned from Southampton to assist in clearing the waterway. Divers attached cables to the *Mary P. Cooper* prior to lifting the wreck on 8 April 1961, nineteen days after the sinking. Another blockage occurred eight years later in 1969, when the *Manchester Courage* ran into and burst open the lock gates at Irlam. (E. Gray)

After the early failure of the Ship Canal Passenger Steamer Company, the few passenger boats seen on the canal were usually Mersey ferry boats, hired for special occasions. The *Bidston* (built in 1903 for the Birkenhead Corporation) is seen here in Mode Wheel Lock in 1927, carrying members of the Association of British Chambers of Commerce. The *Bidston* had participated in the opening ceremony of No. 9 dock in 1905. (Fox Photos)

The Wallasey ferry *Royal Daffodil* carries a full complement as she passes through Barton on a special trip along the canal. Two of the best known of the Mersey ferries, the *Daffodil* and the *Iris*, earned the 'Royal' prefix as a result of their gallant involvement in action at Zeebrugge in 1918. She was taken out of service after twenty-seven years in 1933. The practice of hiring Mersey ferries for excursions along the canal continues to this day.

SECTION EIGHT

SHIP CANAL RAILWAYS

The Manchester Ship Canal Company operated the largest private railway system in the British Isles. At its peak it included over 230 miles of track laid along the length of the canal, and on wharves and sidings by the various docks. The Railway Department owned 75 locomotives, 2,700 wagons, and employed 790 people, the highest total of all occupations working on the canal. The dock railways were able to receive and despatch trains to and from all the main line systems, and, in addition, worked most of the traffic between factories in Trafford Park. The 'gridiron' on Salford Quay was the point where dock railways converged. The locomotive on the right is the *Heywood*, a Manning Wardle tank engine, built in 1888, which was used by the contractor in the construction of the canal, and was afterwards retained to become No. 8 in the Company's fleet. It remained in service until 1927.

The railway swing-bridge in the terminal docks was used for goods trains only. It connected the tracks and sidings in the main docks with the lines on Trafford Wharf and in Trafford Park. In the foreground is the Trafford Wharf coaling stage. In the early years of the canal the sidings opposite, behind No. 6 dock, were the only ones available for marshalling trains. (MSCCo)

The railway bridge was swung to permit the passage of vessels sailing to and from Pomona Docks. Constructed in 1895 by Butler & Company of Leeds, the bridge was single-track only. It weighed 65 tons. It is closed to rail traffic as the Belfast-registered steamer *Clareisland* (built in 1915 for Arthur Guinness) passes through on one of its regular visits to Pomona. (MSCCo)

The railway swing-bridge is crossed by a train heading for Trafford Wharf. The original 1895 bridge served well, but increased traffic in the Second World War, particularly to and from the factories of Trafford Park, was hampered by its single track. Consequently, a new double-track version, weighing 300 tons, was authorized, and was supplied in 1943 by the same firm, Butler & Company. In more recent times, as part of the Salford Quays development scheme, the second bridge was removed and re-installed in a new location as a pedestrian-only footbridge across the former No. 9 dock. (Charles Downs)

The purchase of the former Manchester Racecourse site allowed space for an extensive marshalling yard sited between Nos 8 and 9 docks. No.1 Grid, as it was known, is seen from the roof of the grain elevator in 1935, looking across to No. 8 dock. The main line connection to the Lancashire & Yorkshire line at Windsor Bridge departed via the New Barns Junction, off the picture to the left. (Stewart Bale).

Tracks were laid between quayside and warehouse, enabling direct discharge or loading to or from railway wagons. Hydraulic cranes load export traffic on to the Glasgow steamer *Locksley Hall*, while a Hudswell Clarke locomotive waits at the head of a train on the quay. When a train was fully loaded or unloaded the wagons were drawn out and fresh ones shunted in, such movements usually taking place, for safety reasons, during the dockers' break times. (MSCCo)

A bulk cargo of sulphur is unloaded by grab crane direct into waiting wagons from the Galveston, Texas, steamer *Joseph Lykes*. Such cargoes were unpleasant and dusty to work with, and often justified the payment of extra money to dockers allocated to this task. Carbon black was generally agreed to be the dirtiest cargo, which certainly justified the 'dirty money' addition. (Central Press)

Connections to the main line railways were essential for the collection and distribution of goods. The Cheshire Lines linked on the south side via Trafford Wharf, and the London & North Western line ran via Weaste. The Lancashire & Yorkshire Railway constructed a link from Windsor Bridge on the cut-and-cover principle to New Barns Junction. The title on the signal box reads 'Ship Canal Junction' on this March 1898 picture of the first L&Y engine to enter the docks. (MSCCo)

The Ship Canal Company retained twenty-three locomotives from the construction contract, and from 1897 began to purchase its own engines, which until 1914 were named, usually after ports of the world. Nameplates were removed from the tank sides in 1915 and replaced by numbers, though many engines still displayed their names painted on the sandbox. Number 40, *Buenos Ayres*, draws a cattle train from the Manchester Corporation lairages in about 1930. (MSCCo)

Locomotive no. 53, *Sweden*, was new from Hudswell Clarke in 1911 and served until the end of steam in 1966. The shunter, with his hooked pole for the rapid coupling or uncoupling of wagons, wears his MSC cap as he stands with driver and fireman. (W. Wright)

Trafford Park Estates owned a small number of locomotives for moving wagons between factories, though this work was normally carried out by Ship Canal engines. In 1916 two locomotives ordered from Hudswell Clarke were built to the same design as those supplied to the Ship Canal. Named after a director of the Estates Company, *Lord Ashburton* demonstrates its similarity to *Sweden* above. (Trafford Park Estates Co.)

Ship Canal engines enjoyed sole responsibility for rail workings in Trafford Park after the Estates Company's own locomotives had been sold, though some firms maintained their own locomotives for internal movements. Road/rail crossings, controlled by a flagman at busy junctions, were numerous. Road traffic waits as a train crosses Trafford Park Road in 1964. (E. Gray)

Steam engines began to be replaced by diesel-powered locomotives in 1959. The age of steam was coming to a close in 1964 when no. 31 (*Hamburg*, now preserved on the Keighley & Worth Valley Line) passes the Furness Withy liner *Pacific Stronghold* as she drew a train of wagons across the end of No. 8 dock. (R.L. Wilson)

In June 1966 the last steam locomotive operated on the Ship Canal system. By that date many steam engines had already been withdrawn. This was the melancholy scene at Mode Wheel in April 1966, as locomotiyes 39, 53 and 22 stand partially dismantled, awaiting the further attentions of the oxyacetylene cutter. (E. Gray)

Diesel-hydraulic locomotive no. 32, a Rolls-Royce (Sentinel), suffered an embarrassing mishap when it derailed in its first year of service in July 1966. A total of forty diesel locomotives were purchased between 1959 and 1966, but in 1970 the Company began selling off surplus engines, and by the mid-1980s those few remaining were stationed at Ellesmere Port or Stanlow.

SECTION NINE

THE FURNESS WITHY GROUP

Ship-owner Christopher Furness of West Hartlepool, who, with ship-builder Edward Withy, formed the Furness Withy Company in 1891, was destined to have an important influence on the fortunes of the Manchester Ship Canal. When the canal opened in 1894 the response from major ship-owners was lukewarm, and most saw no reason to depart from their customary schedules. Support for the canal came mainly from north-eastern interests. The first Furness Withy proposal was for a Manchester–Bombay service, to import raw cotton and export finished goods. This scheme was defeated by the combined opposition of established lines, but the threatened competition resulted in both Anchor and Clan Lines offering regular sailings to Manchester. Furness then investigated the Canadian trade, and played a leading role in the formation of Manchester Liners. The Furness Withy enterprise continued to expand, and brought into association many of the best-known shipping lines, whilst retaining their individual identity. James Knott, Newcastle-based owner of the Prince Line, was another north-easterner whose support did much to ensure the success of the Ship Canal. His *Belgian Prince* is about to enter the Mode Wheel Locks on the occasion of the inaugural procession on 1 January 1894. The Prince Line became part of the Furness Withy Group in 1916. (R. Banks)

Sir Christopher Furness (knighted in 1895) urged the establishment of a Manchester-based shipping line, and indicated his willingness to become the major share-holder if local support was forthcoming. The result was Manchester Liners, which began trading with two second-hand ships in 1898. The *Manchester City* was the Company's first new ship. She had extensive refrigerated space for frozen meat, and additional accommodation for 700 live animals. (Norman Edwards Associates)

Manchester Liners traded mainly to North American ports, bringing back imports of cotton from New Orleans or Galveston, or Canadian grain from Montreal. To negotiate the bridges on the canal, the ships had telescopic masts. The *Manchester Corporation* of 1899 discharges grain at the No. 1 grain elevator on Trafford Wharf. (Norman Edwards Associates)

Mode Wheel, 1904. North-eastern interests also promoted the establishment of ship-repair facilities at Manchester. The pontoon dock, described by a Liverpool critic as an 'ugly red monster', had been built on the Tyne and, though rudderless, was successfully towed round the coast to its position by the dry docks. Here it holds James Knott's *Ocean Prince*. The dredger *Mersey* is moored to the left. (Rotary Photographic)

The Prince Line ships traded to Mediterranean ports, exporting the products of Lancashire industry, and bringing back raw cotton, fruit, or general cargo. The *British Prince* approaches Mode Wheel Locks under tow in about 1908. The sender of this postcard was evidently not too happy. He wrote: 'Joined 6th November, left 22nd January – 2 months 17 days!' (Charles Downs)

By the time the *Manchester Exchange* was acquired in 1901, Manchester Liners' fleet had grown to a total of nine ships. By 1904 the fleet numbered fourteen, several of which had been built in the associated Furness Withy shipyards. Services to Boston and Philadelphia were added, and much frozen meat was imported from the Argentine, but after this early period of rapid growth no further ships were ordered until 1912. (Norman Edwards Associates)

The burgeoning factories of Trafford Park exported many of their products via the nearby docks, just as the promoters of the industrial estate had envisaged. Railway carriages built at the American Car & Foundry Company for Egyptian State Railways are loaded on board the *Roman Prince* in Salford Docks in about 1910. (Charles Downs)

In 1914 a Manchester Liner off the coast of north-west Ireland, outward-bound for Montreal, gained the unhappy distinction of being the first merchant ship to be sunk by a mine. Subsequently ten more ships were lost by mine or torpedo, including four hastily acquired replacement vessels. A vessel under construction for Austrian Lloyd at the outbreak of war was diverted to Manchester Liners and became the *Manchester Hero* in 1916. (Charles Downs)

One of six turbine steamers built by the Furness Shipbuilding Company in 1922–3 became the *Manchester Regiment*. At 7,930 gross tons she was Manchester Liners' largest and fastest ship to date. An Atlantic crossing from the Mersey to Quebec was made in seven days nine hours. The derrick framework gave her a characteristic appearance, not admired by all. She was lost in a convoy collision in 1939.

The five ships of the Norfolk and North American Steam Navigation Company were acquired by Furness Withy in 1910. A service to west coast ports of North America via the Panama Canal resulted, and in 1923 the prefix 'Pacific' was adopted for ships sailing this route. The outward-bound *Pacific Shipper* of 1924 leaves the Manchester Ship Canal at Eastham. (Charles Downs)

Between 1924 and 1929 Furness Withy commissioned a total of nine 'Pacific' ships. The inward-bound *Pacific Enterprise* (1928) leaves Barton Locks in 1938, as the stern tug *Acton Grange* waits by the small lock. Approaching (right) is the outward-bound American vessel *City of Alma*. The *Pacific Enterprise*, sailing from Vancouver to Manchester, was wrecked in 1949 when she ran on to a rock in thick fog off the Californian coast. (A.R. Prince)

One of several ships built in the Furness shipyards in the early 1920s for various Group companies was the Prince Line's *Egyptian Prince* (1922), seen in Manchester Docks in 1934, having arrived on the direct service from Alexandria. Her bow plates show evidence of recent repainting, a regular task when in port. She was the fourth ship to bear this name, but not the last, for the name was used yet again for a new vessel in 1951. (Mack & Co.)

The *Cyprian Prince* (third of that name) was built by the Furness yard in 1937. Outward-bound, she demonstrates the limited clearance under the Irlam railway viaduct in 1938–9. Sadly, she had a short career, for she was bombed and sunk by enemy aircraft at Piraeus, Greece, in 1941. (Charles Downs)

On the canal, vessels pass port side to port side, as they do at sea. The inward-bound *Pacific Reliance,* built in 1927 as a sister ship to the *Pacific Enterprise*, has negotiated the narrow point beneath the Runcorn railway viaduct, her masts telescoped for transit to Manchester. The passing ship waits for her to clear the narrows before proceeding. The *Pacific Reliance* was one of five 'Pacific' ships lost in separate attacks by enemy submarines in 1940–3.

Manchester Liners ordered five new vessels in the period 1935–41. One was the *Manchester Progress,* towed by tug *Archer* at Irlam on delivery in 1938. The new ships had automatic stokers, improved crew quarters and cabins for twelve passengers. Some unusual voyages were made in wartime, the *Manchester Progress* being one of the last ships to leave Rangoon in 1941. She survived to remain in service until 1966. (Allied Newspapers)

The *Manchester Shipper* (1943), inward-bound on the canal near Warburton, was an oil-burning turbine vessel. Unusually, she had accommodation for seventy-five passengers as a wartime measure. Her arrival helped compensate for three enemy sinkings. Apart from one trip to the Middle East, she remained on the North Atlantic run until withdrawn in 1969.

Of the eight Furness Withy vessels on the Pacific run in 1939, only three survived the war. American-built 'Liberty' ships, a pre-fabricated design for rapid construction, helped replace lost tonnage in 1942–3. Four were allocated to Furness Withy in 1947, the *Samtredy* being renamed the *Pacific Importer.* In No. 8 dock in about 1950 she was loading cars for export to Vancouver. (Central Press)

The *Pacific Unity* (towed by tug *Arrow* when passing outwards through Barton on her maiden voyage in 1948) was one of six new 'Pacific' vessels built for Furness Withy in post-war years. With a beam of 63 feet 5 inches, she was one of the widest vessels to navigate the canal. There was little room to spare in the 65-feet-wide locks. (Manchester City News)

A high proportion of Manchester Liners' cargo was transhipped at Montreal. When the Canadian Government announced plans for its St Lawrence Seaway project, it was decided to gain a head-start in direct trade to Great Lakes ports by building two ships (referred to as 'Lakers') small enough to negotiate the existing locks on the Lachine, Welland and Soo canals. The *Manchester Pioneer* was launched by Cammell Laird in 1952. (Norman Edwards Associates)

The two original 'Lakers' were joined by a third, purchased from Norwegian owners in 1953. As the direct links to ports such as Toronto and Chicago developed, two more ships were added in 1956. The *Manchester Vanguard* and *Manchester Venture* were motor vessels, with engines and accommodation at the stern. At this time, the familiar black hulls were painted grey. In winter they worked in the Canary Islands fruit trade. The St Lawrence Seaway opened in 1959, rendering the small 'Lakers' redundant. (Norman Edwards Associates)

The Cairn Line, operating a service to Canada from east coast ports of Britain, was another member of the Furness Withy Group. In 1965, as traffic on the route dropped, its three ships were chartered to Manchester Liners' more profitable west coast service. The *Cairnforth* (1959) was renamed *Manchester Freighter*. She lies at the Furness berth in No. 8 dock in 1967.

The *Pacific Stronghold* of 1958, at the customary Furness Withy berthing point at the end of No. 8 dock, was one of six 'Pacific' ships working out of Manchester in the 1960s. All were steam turbine ships, speedy, but with high operating costs when compared to motor vessels. The North Pacific service was closed as a Group economy measure in 1970.

To counter rising shore costs, labour problems, and American competition, Manchester Liners pioneered the change to containerization on its North Atlantic route. The *Manchester Challenge* was the first cellular container ship ordered for a British company. Moored under the special crane at the container terminal by No. 9 dock, she made her first crossing to Montreal in 1968.

Traditional vessels were sold off as new container ships were commissioned, but the *Manchester Quest* was a 1970 conversion of the 1959 *Manchester Miller*. The *Peakdale H* alongside is replenishing the fuel bunkers. The second gantry crane was erected in 1971 as container traffic increased. Ships departed every four or five days. Similar terminal facilities were provided at Montreal, and time spent in port was reduced from as much as ten days to two.

The *Manchester Challenge* outward-bound along the canal is headed by the tug *MSC Sovereign* in this 1974 view from Warburton Bridge. Six and a half days were allowed for the passage from Manchester to Montreal. A year-round service was possible with the aid of an ice-knife fitted round the rudder. This offered protection if the ship had to go astern in the ice. Note the crow's nest look-out post in the bows. (Norman Edwards Associates)

In 1968 Manchester Liners took over the operation of the Prince Line's loss-making Mediterranean services. These services, too, became containerized, a smaller number of vessels carrying twice as much cargo as the conventional ships of earlier years. Loading of containers continued during the hours of darkness. This vessel retained its Prince Line title and its distinctive funnel markings. Furness Withy ships had a black funnel with one narrow and one broad red band; the Prince Line markings were the same, but with the Prince of Wales feathers added. Manchester Liners' funnels, though apparently similar, were red with a broad black top and one black band – that is they did not have the black base of the Furness Withy ships. (Norman Edwards Associates)

SECTION TEN

THE CENTENARY

The centenary of the Manchester Ship Canal was celebrated in 1994. By that date the pattern of the British shipping trade had changed irrevocably. The peak years of the 1950s were followed by a depressed market and fierce competition in the early 1970s. Manchester Liners, by then a wholly owned subsidiary of Furness Withy, traded at a loss. Further troubles were caused by mergers and take-over bids. The Canadian service ceased to operate from Manchester in 1979, and Furness Withy was sold to a Hong Kong-based group in 1980. The Canal Company lost revenue as trade moved elsewhere, container ships grew to a size beyond that which could be accommodated by the canal, and traffic dropped to such an extent that complete closure of the upper reaches of the waterway beyond Runcorn was contemplated. However, management adapted to meet the new conditions, and the western end of the canal continues to be busy with shipping. In the 1980s policy changes transformed the former terminal docklands into an attractive area of mixed residential, business, and leisure interests. Water quality has been improved, and the former docks are now areas for sports, including fishing and boating. Future plans include the construction of an art gallery and a concert hall. The old No. 8 dock (now named Ontario Basin), has an occasional visitor in the shape of a replica of Drake's ship *The Golden Hinde*, moored right. The Quay House (centre) is now a restaurant. Model yachts ply the waters in this view towards the new Welland Lock. Names chosen for the water basins commemorate the former trade connections with North America. (E. Gray)

Moored at Trafford Wharf as a permanent tourist attraction is one of the last conventional minesweepers, *HMS Bronington* (M1115). Introduced in 1953, these ships were of shallow-draught, non-magnetic construction, with wooden hulls and alloy superstructure. The ship, once under the command of Prince Charles, was taken out of service in 1988, and is now owned by the Bronington Trust. (E. Gray)

The western end of the canal remains busy. At Ellesmere Port wharf, on 14 September 1996, the German-registered vessel, *Annika-M* of Hamburg was enlisting the assistance of a dockside crane to lift hatch covers in preparation for loading containers. In the distance may be noted the slopes of the spoil heap 'Mount Manisty'. (N. Jones/Tony Robinson)

Though commercial traffic is no longer seen in the former terminal docks, the dry-docks, now occupied by Lengthline Limited, remain busy. The *Redthorn* (ex-*Pinewood*) a general cargo vessel of the Coe Metcalfe Shipping Company, Liverpool, is under repair in August 1996. Note the anchor chains laid out on the base of the dock. (E. Gray)

The original crossing points on the canal sufficed until the 1960s, when the high-level bridges at Runcorn, Thelwall (M6) and Barton (M63) were constructed. The latest addition, on a new road link between Eccles and Trafford Park, is the 1994 Centenary Bridge, which has a centre lifting section enabling ships to pass through. The *Arklow Viking*, one of several vessels which bring bulk supplies of grain to the Cerestar wharf, sails through in August 1996. (E. Gray)

ACKNOWLEDGEMENTS

A book such as this would not be possible without the work of the photographers who recorded the scenes and events of long ago. The author is grateful to all those people who kindly granted permission for pictures to be reproduced, and the illustrations are individually acknowledged where the photographer is known. As the collection was amassed over half a century, it is regretted that the source of illustrations was not always noted, and apologies are offered for any inadvertent omissions in the credits.

Many of the photographs originated from the Public Relations Department of the Manchester Ship Canal Company. The author's interest in the Canal began in childhood, and a succession of Company officials proved most kind in devoting time to answer questions, supply photographs, grant permits for photography, and generally indulge an amateur's interest in the management and day-to-day workings of the Canal. The Company has always been conscious of the importance of preserving a history of the canal's development, and even before construction began in 1887 photographers were engaged to record scenes which were about to change forever. In a way, this volume is a tribute to those people who financed, documented and preserved the Canal's history from the time of its inception in Victorian times.

The author is particularly grateful to Robert Hough, present Chairman of the Manchester Ship Canal Company, and to Alan Dickinson. Their continued interest has been greatly appreciated, and their permission to reproduce illustrations from Company archives is acknowledged with thanks. In its hundred-year history, the canal has been affected by significant patterns of British trade, industry and transport, but the company has evolved and adapted to meet the new demands and conditions, and now moves into its second century with confidence. The Company's senior executives have encouraged and supported this photographic survey of the Canal's history.

Commercial picture-postcards of the 1900–1925 period have proved a useful source of material. Many reproduced here represent but a small portion of the output of photographers such as Charles Downs, J.L. Brown, A.H. Clarke, T. Pinder and H. Grundy, all local cameramen who visited the docks from time to time, and whose principal sales were often to seamen writing home to their families. From more recent times, the author is particularly grateful to John and Sara Cooper, of Norman Edwards Associates Limited, whose company acted as public relations consultants to Manchester Liners, and John Cooper photographed new vessels and recorded many relevant events from the 1970s onward. Acknowledgement is also made to R.L. Wilson of West Kirby; Bill Newton of Trafford Libraries; staff at Manchester Central Reference Library and Liverpool Maritime Museum; Lengthline Limited, Trafford Wharf; and R. Alexander, formerly of Furness Withy. Alan Palmer of Worsley has drawn the maps with the expertise for which he is noted.

For information used in captions the following works have been consulted:
Burrell, David. *Furness Withy, 1891–1991*, (World Ship Society, 1992)
Corbridge, J. *The Mersey & Irwell Navigation*, (Morton, 1979)
Hallam, W.B. *Tugs of the Manchester Ship Canal*, (MSCCo. 1978)
Leech, Sir Bosdin. *History of the Manchester Ship Canal*, (Sherratt & Hughes 1907)
Stoker, R.B. *The Saga of Manchester Liners*, (Kinglish, Isle of Man, 1984)
Thorpe, Don. *The Railways of the Manchester Ship Canal*, (Oxford, 1984)
Plus issues of *Lloyd's Shipping Register*, *Port of Manchester Guide*, and *Ship Canal Company Regulations*.
The initial idea for compiling this volume came from Simon Fletcher of Sutton Publishing, to whom thanks are extended for his never-failing cheerful encouragement, help, and guidance. To his wife, Kathleen, the author once again records his heartfelt thanks for so many things, not least tolerance, inspiration, advice, and countless cups of coffee.

BRITAIN IN OLD PHOTOGRAPHS

Lincoln
Lincoln Cathedral
The Lincolnshire Coast
Liverpool
Around Llandudno
Around Lochaber
Theatrical London
Around Louth
The Lower Fal Estuary
Lowestoft
Luton
Lympne Airfield
Lytham St Annes
Maidenhead
Around Maidenhead
Around Malvern
Manchester
Manchester Road & Rail
Mansfield
Marlborough: A Second Selection
Marylebone & Paddington
Around Matlock
Melton Mowbray
Around Melksham
The Mendips
Merton & Morden
Middlesbrough
Midsomer Norton & Radstock
Around Mildenhall
Milton Keynes
Minehead
Monmouth & the River Wye
The Nadder Valley
Newark
Around Newark
Newbury
Newport, Isle of Wight
The Norfolk Broads
Norfolk at War
North Fylde
North Lambeth
North Walsham & District
Northallerton
Northampton
Around Norwich
Nottingham 1944–74
The Changing Face of Nottingham
Victorian Nottingham
Nottingham Yesterday & Today
Nuneaton
Around Oakham
Ormskirk & District
Otley & District
Oxford: The University
Oxford Yesterday & Today
Oxfordshire Railways: A Second Selection
Oxfordshire at School
Around Padstow
Pattingham & Wombourne
Penwith
Penzance & Newlyn
Around Pershore
Around Plymouth
Poole
Portsmouth
Poulton-le-Fylde
Preston
Prestwich
Pudsey
Radcliffe
RAF Chivenor
RAF Cosford
RAF Hawkinge
RAF Manston
RAF Manston: A Second Selection
RAF St Mawgan
RAF Tangmere
Ramsgate & Thanet Life
Reading
Reading: A Second Selection
Redditch & the Needle District
Redditch: A Second Selection
Richmond, Surrey
Rickmansworth
Around Ripley
The River Soar
Romney Marsh
Romney Marsh: A Second Selection
Rossendale
Around Rotherham
Rugby
Around Rugeley
Ruislip
Around Ryde
St Albans
St Andrews
Salford
Salisbury
Salisbury: A Second Selection
Salisbury: A Third Selection
Around Salisbury
Sandhurst & Crowthorne
Sandown & Shanklin
Sandwich
Scarborough
Scunthorpe
Seaton, Lyme Regis & Axminster
Around Seaton & Sidmouth
Sedgley & District
The Severn Vale
Sherwood Forest
Shrewsbury
Shrewsbury: A Second Selection
Shropshire Railways
Skegness
Around Skegness
Skipton & the Dales
Around Slough
Smethwick
Somerton & Langport
Southampton
Southend-on-Sea
Southport
Southwark
Southwell
Southwold to Aldeburgh
Stafford
Around Stafford
Staffordshire Railways
Around Staveley
Stepney
Stevenage
The History of Stilton Cheese
Stoke-on-Trent
Stoke Newington
Stonehouse to Painswick
Around Stony Stratford
Around Stony Stratford: A Second Selection
Stowmarket
Streatham
Stroud & the Five Valleys
Stroud & the Five Valleys: A Second Selection
Stroud's Golden Valley
The Stroudwater and Thames & Severn Canals
The Stroudwater and Thames & Severn Canals: A Second Selection
Suffolk at Work
Suffolk at Work: A Second Selection
The Heart of Suffolk
Sunderland
Sutton
Swansea
Swindon: A Third Selection
Swindon: A Fifth Selection
Around Tamworth
Taunton
Around Taunton
Teesdale
Teesdale: A Second Selection
Tenbury Wells
Around Tettenhall & Codshall
Tewkesbury & the Vale of Gloucester
Thame to Watlington
Around Thatcham
Around Thirsk
Thornbury to Berkeley
Tipton
Around Tonbridge
Trowbridge
Around Truro
TT Races
Tunbridge Wells
Tunbridge Wells: A Second Selection
Twickenham
Uley, Dursley & Cam
The Upper Fal
The Upper Tywi Valley
Uxbridge, Hillingdon & Cowley
The Vale of Belvoir
The Vale of Conway
Ventnor
Wakefield
Wallingford
Walsall
Waltham Abbey
Wandsworth at War
Wantage, Faringdon & the Vale Villages
Around Warwick
Weardale
Weardale: A Second Selection
Wednesbury
Wells
Welshpool
West Bromwich
West Wight
Weston-super-Mare
Around Weston-super-Mare
Weymouth & Portland
Around Wheatley
Around Whetstone
Whitchurch to Market Drayton
Around Whitstable
Wigton & the Solway Plain
Willesden
Around Wilton
Wimbledon
Around Windsor
Wingham, Addisham & Littlebourne
Wisbech
Witham & District
Witney
Around Witney
The Witney District
Wokingham
Around Woodbridge
Around Woodstock
Woolwich
Woolwich Royal Arsenal
Around Wootton Bassett, Cricklade & Purton
Worcester
Worcester in a Day
Around Worcester
Worcestershire at Work
Around Worthing
Wotton-under-Edge to Chipping Sodbury
Wymondham & Attleborough
The Yorkshire Wolds

To order any of these titles please telephone our distributor, Littlehampton Book Services on 01903 721596
For a catalogue of these and our other titles please ring Regina Schinner on 01453 731114